T0305504

Smart and Sustainable Operations Management in the Aviation Industry

The ongoing impact of Industry 4.0 and disruptive technologies has transformed conventional supply chains into globally connected collaborative networks. As supply chains turn into flexible, agile, and digital structures, the planning and operation phases of the key business processes become more complex. This book presents state-of-art chapters on smart and sustainable supply chain management in the aviation industry.

The aviation industry is one of the main industries affected by the rapid transformation initiated by the Fourth Industrial Revolution. Disruptive technologies such as artificial intelligence, augmented reality, advanced sensor systems, and autonomous robots, are shaping the future of the aviation industry. In this sense, conventional aviation operations are being replaced by innovative methods, and aviation companies need to switch their business models. This transformation is required due to the increasing economic, environmental, and social concerns.

This book covers a wide range of topics, including key business operations in aviation, productivity improvement strategies in the aviation industry, and promising applications of disruptive technologies for aviation companies.

Smart and Sustainable Operations and Supply Chain Management

Series Editor: Turan Paksoy

This new book series will focus on smart applications of Industry 4.0 and how that involves operations and supply chain management and the triple-bottom line approach of: 1) Economic impact, by increasing productivity and achieving operational excellence; 2) Social Impact, by generating human-centric workplaces; 3) Environment impact, by eliminating waste and reducing material usage and applying recycling and reuse in the product design process. The series will include topics such as product price, quality, cooperation, reliability, on time delivery, lean production, corporate social responsibility, green production, green transportation and green packaging for smart and sustainable operations/supply chain management.

Smart and Sustainable Operations and Supply Chain Management in Industry 4.0
Edited by Turan Paksoy and Muhammet Deveci

Smart and Sustainable Operations Management in the Aviation Industry
A Supply Chain 4.0 Perspective
Edited by Turan Paksoy and Sercan Demir

Smart and Sustainable Operations Management in the Aviation Industry
A Supply Chain 4.0 Perspective

Edited by
Turan Paksoy
Sercan Demir

CRC Press
Taylor & Francis Group
Boca Raton London New York

CRC Press is an imprint of the
Taylor & Francis Group, an **informa** business

Designed cover image: Shutterstock – stock photo mania

First edition published 2025
by CRC Press
2385 NW Executive Center Drive, Suite 320, Boca Raton FL 33431

and by CRC Press
4 Park Square, Milton Park, Abingdon, Oxon, OX14 4RN

CRC Press is an imprint of Taylor & Francis Group, LLC

ISBN: 978-1-032-48154-8 (hbk)
ISBN: 978-1-032-48456-3 (pbk)
ISBN: 978-1-003-38918-7 (ebk)

DOI: 10.1201/9781003389187

Typeset in Times
by SPi Technologies India Pvt Ltd (Straive)

Contents

Preface

Especially after the second half of the 90s, the Internet became widespread, and the developments in computer and software technologies were accepted as the beginning of a new era with the introduction of Industry 4.0 in 2011. The aviation industry, which has used edge technologies the most since World War II, is experiencing a significant transformation with Industry 4.0 technologies. Automation, artificial intelligence, and sensor technologies are used in a wide range of areas, from aircraft production to maintenance, to increase operational efficiency and reduce costs. These innovations also increase flight safety and raise standards in the industry.

On the other hand, in parallel with the rise of smart technologies, the world has begun to feel the effects of the climate crisis more, and societies and lawmakers have begun to take more conscious and concrete steps toward sustainability. This required them to see the concept of sustainability as an indispensable instrument for businesses, no matter what they do. Fortunately, these two concepts do not compete; on the contrary, they feed and grow each other. The smart systems brought by Industry 4.0 increase fuel efficiency and reduce carbon footprint. Technological developments in aircraft engines, especially in the aviation sector, support sustainability by encouraging the integration of green fuels and hybrid technologies.

In this new era called Industry 4.0, with the advantage of being in a technology-friendly and leading industry, airline companies are moving towards a greener future by adopting solutions such as smart route planning and using sustainable fuels to minimize environmental impacts. Data Analytics, perhaps one of the most popular Industry 4.0 technologies, enables more effective evaluation of flight data and maintenance records in the aviation industry. The Internet of Things (IoT) ensures sustainable maintenance and performance optimization by ensuring communication between sensors and components on aircraft. Artificial Intelligence (AI) and Machine Learning (ML) optimize energy efficiency while improving safety in flight simulations, pilot training, and air traffic management. 3D printing technology enables effective use of resources, reduces waste for spare parts production, and allows rapid prototyping. Robotic applications, which we have begun to witness, especially in airports, increase customer satisfaction by making passengers' work easier, supporting technical personnel in maintenance and repair processes, ensuring human safety in challenging and risky tasks, and accelerating business processes. Similarly, Blockchain Technology is a factor that increases sustainability in supply chain management in the aviation industry regarding parts tracking and reliability. On the other hand, Digital Twin, another Industry 4.0 technology, is used for real-time monitoring and maintenance of aircraft components, detecting possible malfunctions, and developing solutions to eliminate these malfunctions. Industry 4.0 also reduces fuel consumption and carbon emissions by making air traffic management more effective and safe.

This book discusses Industry 4.0 technologies that support sustainability and presents examples of their use in the aviation industry. We hope it will be helpful for academic fellows, students, and practitioners in the sector, and we wish you a pleasant reading.

Turan Paksoy and Sercan Demir

About the Editors

Dr. Turan Paksoy received his B.S. degree in Industrial Engineering from Gazi University in 1998. He earned his M.S. degree in Industrial Engineering in 2001 and Ph.D. degree in Production Management from Selçuk University in 2004. He received an associate degree in production management in 2008 from the National Academic Council of Turkey. In 2015, he was promoted to full professor at Selçuk University. Then, Dr. Paksoy attended London Kingston University as a visiting professor in 2015. Dr. Paksoy worked as a faculty member at Konya Technical University, Department of Industrial Engineering, from June 2018 to October 2020. He was assigned as Dean of the Faculty of Aviation and Space Sciences at Necmettin Erbakan University in November 2020. He served 2.5 years as Dean and offered courses at the Aviation Management Department till September 2023 at this faculty.

Dr. Paksoy has been working as a faculty member at Necmettin Erbakan University, Department of Industrial Engineering since September 2023. He was awarded the position of Visiting Professor by the Faculty of Business and Law Northumbria University Newcastle U.K. from January 2024-January 2027. He was honoured to be chosen for "Global Sustainable Leaders" by the U.S Green Chamber of Commerce in 2018. Dr. Paksoy has over 300 published papers (45 in SCI-indexed journals). His papers appear frequently in reputable and internationally recognized journals such as Computers and Industrial Engineering, the European Journal of Operational Research, Transportation Research, and the International Journal of Production Research. Dr. Paksoy has three edited books titled "Lean & Green Supply Chain Management", "Logistics 4.0: Digital Transformation of the Supply Chain Management", and "Smart & Sustainable Operations and Supply Chain Management in Industry 4.0" published by Springer and Taylor & Francis CRC Press in Germany and the United States.

Dr. Sercan Demir received his B.S. degree in Industrial Engineering from Selcuk University in 2008. He completed his M.S. degree in Industrial and Systems Engineering at the University of Southern California, Los Angeles in 2012. He was awarded his Ph.D. in Industrial Engineering from the University of Miami in 2017. Dr. Demir's research focuses on Industry 4.0 and Digital Transformation, Sustainable SCM, Disaster Risk Management, Humanitarian SCM, and Game Theory Applications in SCM. His papers appear frequently in reputable and internationally recognized journals such as *Computers & Industrial Engineering, Journal of Cleaner Production, Socio-Economic Planning Sciences*, and *Expert Systems with Applications*. In addition to his extensive research contributions, Dr. Demir has authored chapters in two influential books: *Logistics 4.0: Digital Transformation of Supply Chain Management*, and *Smart and Sustainable Operations and Supply Chain Management in Industry 4.0*. His chapters provide valuable insights into the evolving landscape of supply chain management in the digital age, addressing the challenges and opportunities presented by Logistics 4.0 and the integration of smart and sustainable practices in Industry 4.0.

Contributors

Kemal Alaykiran
Necmettin Erbakan University
Konya, Turkiye

Batin Latif Aylak
Faculty of Engineering, Department
of Industrial Engineering
Turkish-German University
Istanbul, Beykoz, Turkey

Celal Cakiroglu
Turkish-German University
Istanbul, Turkey

Engin Hasan Çopur
Necmettin Erbakan University
Meram, Konya, Turkey

Sercan Demir
Department of Marketing, Operations
and Systems, Newcastle Business
School
Northumbria University
Newcastle upon Tyne, United Kingdom

Nurcan Deniz
Eskisehir Osmangazi University
Eskisehir, Türkiye

Beyzanur Cayir Ervural
Department of Aviation Management,
Faculty of Aviation and Space
Sciences
Necmettin Erbakan University
Konya, Meram, Turkey

Mehmet Akif Gunduz
Necmettin Erbakan University
Meram, Konya, Turkey

Basar Koc
Stetson University
DeLand, Florida

Mehmet Hakan Özdemir
Faculty of Economic and Administrative
Sciences, Department of Business
Administration
Turkish-German University
Istanbul, Turkey

Turan Paksoy
Necmettin Erbakan University
Meram, Konya, Turkey

Büşra Yiğitol
Necmettin Erbakan University
Konya, Türkiye

1 A Conceptual Framework for Smart and Sustainable Operations in the Aviation Industry

Sercan Demir
Northumbria University, Newcastle upon Tyne,
United Kingdom

Turan Paksoy
Necmettin Erbakan University, Konya, Turkey

1.1 INTRODUCTION

The aviation industry covers nearly all aspects of air travel and the operations and activities that support and contribute to it. For instance, airline and air travel industries, aircraft manufacturing, and military aviation are parts of it. The aviation industry is growing rapidly and will continue to do so exponentially in the following decades. Furthermore, airline companies and aircraft manufacturers are engaged in fierce competition while trying to maintain sustainable operations. The aviation industry therefore faces unprecedented challenges of meeting passenger growth and service demand quality while reducing carbon emissions.

Industry 4.0 has brought about smart tools and technologies, such as artificial intelligence, advanced robotics, augmented and virtual reality solutions, machine learning, big data analytics, cloud systems, and blockchain, which support efficient and sustainable growth in the aviation industry. These technologies present the aviation industry with unmissable opportunities, and their applications support the growth of secure, sustainable, and effective aviation operations. The implementation of one or more smart technologies in the aviation sector helps companies achieve operational efficiency while keeping carbon emissions at a lower level. Hence, a company becomes more flexible, cost-conscious, and sufficiently resilient to large-scale disruptions such as global pandemics, airspace closures, and government restrictions. This chapter introduces smartness and sustainability concepts in terms of aviation perspective and investigates the application of smart tools and sustainability ideas in the aviation sector.

DOI: 10.1201/9781003389187-1

1

1.2 OPERATIONS MANAGEMENT

Operations Management (OM) includes management techniques and business practices concerned with creating better goods and services by improving operational efficiency while increasing a company's competitiveness level in the market (Demir & Paksoy, 2023). According to Heizer et al. (2017), OM techniques apply to any system that produces goods or services. The authors define OM as "the set of activities that creates value in the form of goods and services by transforming inputs into outputs." An effective application of the tools and techniques offered by OM is required to maintain the efficient production of goods and services. Since the start of the Industrial Revolution, the OM tool and methodologies have been developing continuously, pushing businesses to modernize their processes, operations, and corporate cultures in order to keep up with the new trends and increase operational effectiveness (Demir & Paksoy, 2023).

Operations management is a crucial field that deals with the design, planning, execution, and control of business operations to achieve organizational goals efficiently and effectively. The primary objective of operations management is to streamline the production process and enhance productivity, quality, and profitability, while reducing costs and waste. This field aims to optimize the use of resources, including labour, equipment, materials, and time, to meet customer demands and expectations. It also involves ensuring that the production process complies with legal and regulatory requirements, promoting a safe and healthy work environment, and developing strategies for continuous improvement. Overall, operations management aims to create a sustainable, competitive advantage for the organization in the long run.

The market previously operated on the basis of individual needs due to insufficient production capacity. As a result, businesses started to concentrate on producing a large number of standardized products as the market for consumer goods expanded. Customers started to demand customized products once their basic needs were fulfilled. Companies adopted production systems that could produce a medium variety and volume of goods to meet this demand. In recent decades, customer demand has shifted from standardized to customized as the market has expanded from local to global (Gunasekaran & Ngai, 2012).

Globalization currently has a significant impact on business. Since its emergence, the field of operations management has undergone constant change and expanded to include topics that broaden the scope of operations functions. Determining OM boundaries is still a challenging task. Through the use of emerging technologies, initiatives to integrate organizational business and supply chain functions over the past ten years may be expanded into integrated e-business applications. It is anticipated that OM models will play a significant role in the manufacturing and service industries with the integration of e-business applications. Besides, the development of the OM field may be influenced by key enabling technologies, arising customer demand resulting from technological advancements, and new performance measures, in addition to speed, timeliness, and responsiveness (Bayraktar et al., 2007).

TABLE 1.1
The Evolution of Operations Management

Period	Objectives	Strategies/Technologies
Beginning	Individual customer requirements	Craftsman production, artesian production
Post-World War II	Immense demand for consumer products	TQM, JIT, transfer line production systems
1975–1985	Medium volume and medium variety	QRM, CIM, FMS, and BPR
1985–1995	Cost reduction, high variety, and low volume	Lean, agile, and physically distributed enterprise environments
1995–2010	Higher variety and very low volume	Outsourcing, global manufacturing and market, agile, Internet-enabled SCM, 3PL
2010–	Global individualized products and services	Craftsman production, artesian production

Source: Adapted from Gunasekaran and Ngai (2012).

The development of OM is evaluated under four stages of supply chain management: productivity and competitive strategies, physical inflow of materials, production planning and control, and physical outflow of materials (Gunasekaran & Ngai, 2012). Table 1.1 shows the evolution of OM.

Operations management is vital in today's business environment, where companies face fierce competition, rapid technological advancements, and changing customer preferences. Effective operations management enables businesses to optimize their resources and improve efficiency, thereby reducing costs and enhancing productivity. It also helps organizations to adapt quickly to changing market conditions, improve customer satisfaction, and gain a competitive advantage.

However, operations management also faces several challenges in today's business environment. The increasing complexity of supply chains and logistics networks presents crucial challenges. Due to the geographically dispersed business formation, managing a supply chain has become a complex task, involving coordination with multiple stakeholders across different countries and time zones. The need to balance cost-effectiveness with sustainable practices is another key issue. Companies must find ways to reduce environmental impact while retaining profitable operations, which requires innovative solutions and strategic planning. Furthermore, technology is constantly evolving, and businesses must keep up with the latest trends to remain competitive. The rise of automation, artificial intelligence, and the Internet of Things (IoT) presents both opportunities and challenges for operations management. Adopting new technologies can streamline processes, reduce costs, and improve efficiency, but it also requires significant investment and specialized skills. Overall, operations management is critical in navigating the challenges of the modern business environment, and organizations that prioritize it are more likely to succeed in the long run.

Companies benefit from effective operations management practices in several ways:

1. **Efficiency and Cost Reduction**: Effective operations management ensures streamlined processes, smooth production flow, and efficient utilization of resources, leading to cost reduction. Companies can minimize waste, improve productivity, and lower operational expenses by optimizing production, inventory flow, and supply chain processes.
2. **Quality Control**: Operations management focuses on maintaining and improving product or service quality. By implementing quality control measures, companies can minimize defects, enhance customer satisfaction, and reduce the need for rework or returns. This, in turn, helps build a strong reputation, a high customer retention rate, and brand loyalty.
3. **Timely Delivery**: Efficient operations management helps companies meet customer demands on time. By optimizing production schedules, inventory levels, and logistics operations, companies ensure the timely delivery of products or services, leading to high customer satisfaction and profitability.
4. **Flexibility and Adaptability**: Effective operations management allows companies to rapidly adapt to changing market conditions and customer needs. Companies can quickly transform their operations to seize opportunities in the market and mitigate supply chain risks by adopting agile, flexible, and responsive supply chain practices.
5. **Competitive Advantage**: Companies can gain a competitive edge in the market by implementing effective operations management strategies. Successful application of operations management enables companies to deliver products or services at competitive prices while maintaining quality and efficiency. As a result, companies attract more customers, increase their market share, and outperform their competitors.
6. **Innovation and Continuous Improvement**: Operations management involves continuous evaluation and improvement of processes. By fostering a culture of innovation and continuous improvement, companies can identify opportunities for optimization, implement new technologies or techniques, and stay ahead of the competition.
7. **Risk Management**: Operations management includes assessing and managing production, supply chain, and logistics risks. Companies can minimize disruptions and ensure the continuity of their business operations by identifying potential risks and implementing risk mitigation strategies.
8. **Scalability**: An efficient operations management system provides a foundation for scalability. As a company grows, efficient operations allow for seamless expansion, increased production capacity, and the ability to serve a more extensive customer base without compromising quality or increasing costs significantly.

In summary, operations management enables companies to achieve more efficient operations, cost reduction, better quality control, dependable and cost-effective delivery, competitive advantage, innovation and adaptability capabilities, risk management ability, and scalability skills. These benefits increase profitability, customer satisfaction, and overall business success.

1.3 SMART AND SUSTAINABLE OPERATIONS MANAGEMENT

Smart and sustainable operations management refers to implementing smart and environmentally friendly practices in managing the operations of a company. This concept integrates advanced technologies, data analytics, and sustainable principles to optimize business processes, reduce ecological impact, and achieve long-term sustainability goals. Some key aspects of smart and sustainable operations management are:

1. **Technology Integration**: Smart operations management leverages technologies such as the Internet of Things (IoT), advanced robotics, artificial intelligence (AI), and data analytics. These technologies enable real-time monitoring, data-driven decision-making, predictive maintenance, and process optimization. By leveraging these technologies, companies can improve operational efficiency, reduce errors, and enhance resource utilization.

2. **Energy Efficiency**: Sustainable operations management focuses on minimizing energy consumption and maximizing energy efficiency. This involves implementing energy-efficient equipment, optimizing production processes to reduce energy waste, and adopting renewable energy sources. Energy monitoring systems and smart grids can help track energy usage and identify areas for improvement.

3. **Waste Reduction and Recycling**: Sustainable operations management emphasizes waste reduction and recycling efforts. Companies can implement waste management systems to minimize waste generation, optimize recycling processes, and promote the use of recycled materials. By reducing waste, companies can minimize their environmental footprint and potentially save on costs.

4. **Supply Chain Sustainability**: Smart and sustainable operations management can be extended to all supply chain operations. Companies can collaborate with suppliers to ensure ethical sourcing, reduce carbon emissions in transportation, and optimize inventory management to minimize waste. Implementing sustainable procurement practices and working with suppliers prioritizing environmental and social responsibility are crucial for sustainable operations.

5. **Circular Economy Practices**: Smart and sustainable operations management embraces the principles of a circular economy by targeting to eliminate waste and encouraging the reuse of resources. This involves designing recyclable products, implementing reverse logistics systems for product take-back and refurbishment, and exploring opportunities for remanufacturing or repurposing.

6. **Environmental Monitoring and Compliance**: Smart and sustainable operations management incorporates monitoring and complying with

environmental regulations. This practice includes tracking emissions, wastewater management, and committing to environmental standards. Companies can use environmental management systems such as ISO 14001 to ensure compliance and drive continuous improvement in environmental performance.

7. **Stakeholder Engagement**: Smart and sustainable operations management involves engaging with stakeholders, including employees, customers, and communities. By promoting sustainability initiatives internally and externally, companies can foster a culture of responsibility and gain support from stakeholders. This engagement can lead to innovative ideas, increased employee morale, and improved brand image.

By integrating smart technologies and sustainable practices, companies can achieve operational efficiency, reduce environmental impact, and contribute to a more sustainable future. Smart and sustainable operations management benefits the environment and can lead to cost savings, increased competitiveness, and improved brand value. Smart operations management is enabled by smart technologies, as shown in Figure 1.1. These smart technologies are summarized in Table 1.2.

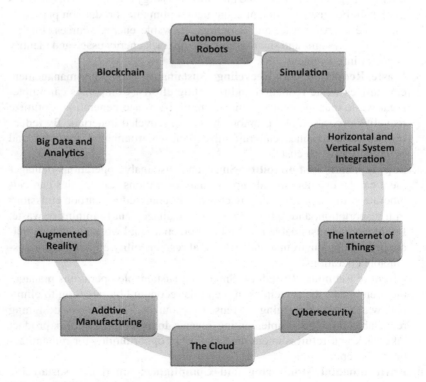

FIGURE 1.1   Smart technologies reshaping the production process (Demir et al., 2020).

TABLE 1.2

The Nine Technologies that are Reshaping the Production

Advanced robots	• Autonomous, cooperating industrial robots with integrated sensors and standardized interfaces
Additive Manufacturing	• 3D printers, used predominantly to make spare parts and prototypes • Decentralized 3D printing facilities, which reduce transport distances and inventory
Augmented Reality	• Digital enhancement, which facilities maintenance, logistics, and SOPs - • Display devices, such as glasses
Simulation	• Network simulation and optimization, which use real-time data from intelligent systems
Horizontal and vertical system integration	• Data integration within and across companies using a standard data transfer protocol • A fully integrated value chain (from supplier to customer) and organization structure (from management to shop floor)
The Industrial Internet of Things	• A network of machines and products • Multidirectional communications among networked objects
Cloud Computing	• The management of huge volumes of data in open systems • Real-time communication for production systems
Cyber Security	• The management of heightened security risks due to a high level of networking among intelligent machines, products, and systems
Big data and analytics	• The comprehensive evaluation of available data (from CRM, ERP, and SCM systems, for example, as well as from an MES and machines) • Support for optimized real-time decision-making

Source: Adapted from Brunelli et al. (2017).

1.4 SMART AND SUSTAINABLE APPLICATIONS IN AVIATION

Airports are at the forefront of technological advancement, primarily because of the exponential annual growth in air travel passengers (Lykou et al., 2018). Airports are investing in technology for a variety of reasons. The primary reasons are to improve operational and cost efficiency (Halpern et al., 2019). Airports have been constantly transforming and implementing digital technologies to increase efficiency in operations, improve passenger experience, generate ancillary revenues, and increase the capacity of existing infrastructure (Tan & Masood, 2021).

The aviation industry has made significant moves in recent years towards adopting smart and sustainable technologies. Due to the process, as well as the labour- and capital-intensive nature of airline operations, even small changes in operations that reduce waste significantly impact the quality of customer service, employee productivity, and overall operations efficiency (Demir & Paksoy, 2021). According to Rajapaksha and Jayasuriya (2020), future airport operations will be based on the "smart airport" concept, which could fundamentally alter how the aviation industry adopts new technologies. The fourth industrial revolution has caused the smart airport concept to advance globally, eliminating the shortcomings of the traditional airport system.

Many researchers and practitioners have already begun to investigate the use of smart technologies and sustainable practices in the aviation industry. Dutta et al. (2020) assert that the integration of blockchain technology with RFID and IoT has the potential to transform the industry by improving port tracking and scheduling of air logistics and transportation. Airbus is one of the earliest companies to integrate its supply chain operations with RFID. Elhmoud and Kutty (2021) conducted a literature review to investigate several models and tools used for sustainability assessment in the aviation industry. In it, they state that it is vital to comprehend the economics of sustainability and eco-efficiency to achieve sustainable airline operations. Halpern et al. (2019) employ the idea of maturity models for airport digital transformation, primarily from the passenger experience perspective, by aiming to create a conceptual framework for the essential aspects of airport digital transformation and digital maturity. Lykou et al. (2018) discuss how IoT technologies and technological advancements may alter aviation security threat models and affect the operational efficiency of smart airports. They investigate the rate of cybersecurity measures being implemented in commercial airports, malicious threats that emerge as a result of installed IoT and smart devices, and risk scenario analysis for IoT malicious attacks with threat mitigation measures.

Michalski et al. (2020) state that it is essential to continuously strengthen and improve airport security management and security systems so that they can handle new challenges. This is because threats are becoming more complex, and terrorist attack planners are using increasingly sophisticated methods and techniques. The authors introduce the concept and general characteristics of smart, selective, and personalized security control systems. Later, they structure the analytical field, analyze the problem in terms of its implementation conditions, opportunities, threats, potential for conflict, and controversies, and suggest future directions for further research. Rajapaksha and Jayasuriya (2020) focus on the idea of smart airports and identify the advantages of smart airport implementation under four areas: aviation security, passenger convenience, operational efficiency, and optimization of limited resources. Some smart airport applications are smart check-in, self-boarding, indoor navigation, biometric services, smart wearables, RFID baggage tags, self-baggage tagging, kiosks for lost luggage, border control, and airport apps for mobile devices.

Zaharia and Pietreanu (2018) discuss trends in airport digitization, the framework for implementing total airport management, and the improvements to airport management that are triggered by new implementation schemes. Price et al. (2013) present the organizational challenges and opportunities associated with adopting design-led innovation within a top Australian airport corporation. They examine the organizational difficulties and roadblocks encountered by a leading Australian airport corporation as it made efforts to incorporate design as a strategic capability. Koroniotis et al. (2020) present a comprehensive analysis of the IoT-enabled smart airport applications and services. They discuss various forms of cyber defence tools, such as AI and data mining methods, and evaluate their advantages and disadvantages in relation to smart airports. Additionally, they classify smart airport subsystems according to their function and level of importance and discuss cyber threats that could jeopardize smart airport's network security.

Besides airport operations, smart and sustainable practices have been implemented in other areas of the aviation industry. One of the key areas of focus has been on improving aircraft design and manufacturing processes. Advanced materials such as carbon composites are increasingly used to design and build lighter and more fuel-efficient aircraft, reducing carbon emissions and overall environmental impact. Additionally, the integration of smart sensors and monitoring systems in aircraft allows for real-time data collection, enabling predictive maintenance and optimizing fuel consumption, leading to cost savings and reduced downtime. The development of alternative fuels and propulsion systems is a vital aspect of sustainable technology use in the aviation industry. For instance, electric and hybrid-electric propulsion systems are being explored for smaller aircraft, contributing to quieter and cleaner operations, particularly in urban air mobility scenarios. As battery technology improves, the feasibility of electric propulsion for larger commercial planes may become a reality, revolutionizing the way we fly and making aviation even more sustainable in the future.

In recent years, the aviation sector has started to utilize digital technology and data analytics to streamline operations and improve air traffic management. Airlines and air traffic control authorities can better predict and manage congestion by utilizing advanced algorithms and artificial intelligence, which results in more direct flight paths, shorter flight times, and less fuel consumption. Additionally, smart technology helps passengers by enhancing convenience and reducing paper waste through personalized travel experiences, streamlined check-ins, and digital boarding passes. Adopting smart and sustainable technologies leads to greener and more efficient processes. From innovative aircraft designs and materials to alternative fuels and digital optimization, these advancements are not only driving environmental conservation, but are also improving operational efficiency and enhancing the overall passenger experience. As the industry continues to invest in and embrace these transformative technologies in the coming years, air travel will become more sustainable and eco-friendly.

1.5   CASE STUDY: SMART AND SUSTAINABLE APPLICATIONS IN AVIATION INDUSTRY

The aviation industry is under increasing pressure to reduce its environmental impact. Smart technologies and sustainable practices offer a promising way to make air travel more efficient and sustainable. Smart technologies can be used to improve the efficiency of aircraft, airports, and air traffic control. For example, predictive analytics can be used to optimize flight paths and reduce fuel consumption. Drones can be used to inspect aircraft for damage and perform maintenance tasks, reducing the need for ground crews. Artificial intelligence can be used to develop new air traffic control systems that are more efficient and less disruptive to the environment. Besides, sustainable practices can help to reduce the environmental impact of air travel. For instance, airlines can use sustainable fuels, such as biofuels and hydrogen, leading them to offset their emissions and reduce greenhouse gases. By combining smart technologies and sustainable practices, the aviation industry can make air travel more sustainable and help to protect the environment.

Like many other airlines and companies in the aviation industry, Lufthansa has been working on various sustainable operation goals to minimize their environmental impact. The company states that sustainable and responsible entrepreneurial practices are the key components of the corporate strategy. The company is dedicated to generating added value for its customers, staff, and investors and meeting its responsibilities towards the environment and society. Due to these factors, Lufthansa continuously enhances its efforts to protect the environment and the climate, uphold fair and responsible employee relations, and actively participate in several social issues (Lufthansa Group, 2022). According to Lufthansa Group Sustainability 2022 Fact Sheet (2022), Lufthansa aims to cut its net carbon emissions from flight operations in half by 2030 compared to 2019 and to become carbon-neutral by 2050. Additionally, the company aims to be carbon-neutral on ground operations in its home markets by 2030. Figures 1.2 and 1.3 depict Lufthansa Group's climate goals.

Turkish Airlines, a company devoted to sustainable development, has created a sustainability agenda built on the four pillars of governance, economy, environmental protection, and social responsibility. The company works with business partners, suppliers, NGOs, and academic institutions on material sustainability issues to overcome obstacles and improve performance (Turkish Airlines – Investor Relations, 2023).

The primary focus of Turkish Airlines' sustainability strategy is the social, economic, and environmental problems that could impact or be impacted by the Incorporation's products, services, or activities. The roadmap for Turkish Airlines' sustainability strategy is developed by considering the most important social, economic, and environmental issues, the corporation's mission, vision, and core values, long-term objectives, and potential risks and opportunities. The Incorporation's sustainability strategy, which is managed using a dynamic approach, is regularly reviewed in light of the findings of the materiality analysis, any new regulations that have been put on the agenda recently, and input from stakeholders (Turkish Airlines – 2021 Sustainability Report, 2021) (Figure 1.4).

Turkish Airlines is engaged in digital transformation efforts to switch many human-made business processes to robotic systems and direct the current workforce towards tasks where they can add more value. The company has carried out many

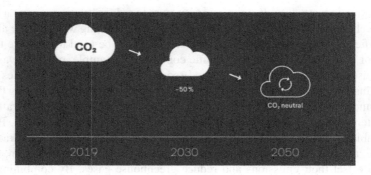

FIGURE 1.2 Lufthansa Group climate action goals (Lufthansa Group Sustainability 2022 Fact Sheet, 2022).

THE LUFTHANSA GROUP'S TOOLS FOR ACHIEVEMENT OF ITS CLIMATE GOALS

Fleet modernisation

More modern and efficient aircraft provide the greatest leverage for reducing CO_2 emissions in this decade.

Efficiency in flight operations

Tools used every day to reduce carbon emissions include intelligent route planning, modern approach procedures and the latest technologies.

Offsetting

High-quality, certified offset projects that promote climate change mitigation worldwide are complemented by CORSIA, the international offset instrument for carbon-neutral growth in air transport.

Sustainable fuels

The key to making flying more climate-friendly is to increase the use of sustainable fuels.

Alternative transport to hubs

The expansion and interconnection of air, rail and bus services makes it possible to reduce the number of short-haul flights and offer alternative modes of travel to flights.

FIGURE 1.3 Lufthansa Group's tools for achievement of its climate goals (Lufthansa Group Sustainability 2022 Fact Sheet, 2022).

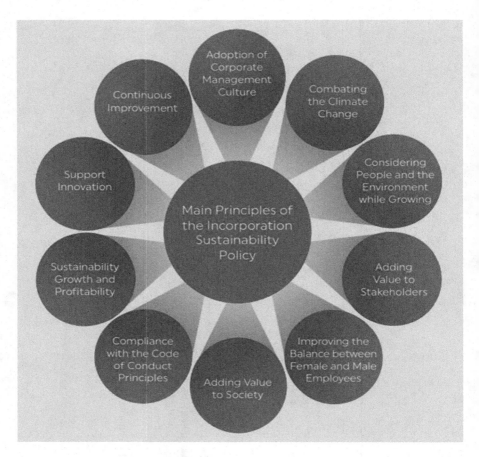

FIGURE 1.4 Main principles of Turkish Airlines Sustainability Policy (Turkish Airlines – 2021 Sustainability Report, 2021).

projects to modernize current systems or to digitize business processes. One of these efforts, the Robotic Process Automation Project, has made it possible for robots to perform repetitive tasks in 15 different business processes, many of which do not require decisions that are based on human intuition and experience. The company wants to build a system infrastructure with the Cargo Work Order Management Project that will make it possible for cargo business processes to use labour and space more effectively. The Horizon Interactive Awards-winning Flight Tracker application, on the other hand, is a touch sphere that successfully displays the extensive flight network and allows for real-time tracking of the aircraft. The company aims to give passengers a closer look at Turkish Airlines' flight network with this application, which will be placed in the private passenger lounge of the Istanbul Airport international terminal (Turkish Airlines – 2021 Sustainability Report, 2021).

In addition to these initiatives, numerous projects are being undertaken to enhance and digitize all aspects of the travel experience. A travel assistant powered by artificial intelligence, airport maps, contactless check-in and boarding applications at the airport, biometric and rapid transit systems, automatic baggage delivery systems, interactive in-flight screen applications, in-flight internet service, and audio and image processing applications are some of the projects included in this list (Turkish Airlines – 2021 Sustainability Report, 2021).

1.6 CONCLUSION

In conclusion, the implementation of smart and sustainable operations in the aviation industry stands as a critical step towards reshaping the future of aviation. The convergence of smart technologies and sustainable practices has introduced a new era of efficiency, safety, and environmental responsibility. As the aviation industry grows, strategically incorporating new technologies like AI-driven predictive maintenance, cutting-edge air traffic management systems, and fuel-efficient aircraft design will streamline operations and reduce environmental impact. Moreover, the aviation industry can reduce its environmental impact and significantly contribute to international efforts to curb climate change by adopting sustainable practices like using alternative fuels, carbon offset programmes, and eco-friendly ground operations. Eventually, the union of technological development and ecological awareness shall improve the aviation industry's competitiveness and pave the way for a greener, smarter, and more sustainable future for aviation.

REFERENCES

Bayraktar, E., Jothishankar, M. C., Tatoglu, E., & Wu, T. (2007). Evolution of operations management: past, present and future. *Management Research News*.

Brunelli, J., Lukic, V., Milon, T., & Tantardini, M. (2017). Five lessons from the frontlines of Industry 4.0 [Electronic resource]. BCG–Mode of access: http://image-src.bcg.com/Images/BCG-Five-Lessons-from-the-Frontlines-of-Industry-4.0-Nov-2017_tcm9-175989.pdf

Demir, S., & Paksoy, T. (2021). Lean management tools in aviation industry: New wine into old wineskins. *International Journal of Aeronautics and Astronautics*, 2(3), 77–83.

Demir, S., & Paksoy, T. (2023). Fundamental concepts of smart and sustainable operations and supply chain management. In *Smart and Sustainable Operations and Supply Chain Management in Industry 4.0* (pp. 1–26). CRC Press.

Demir, S., Paksoy, T., & Kochan, C. G. (2020). A conceptual framework for Industry 4.0: (How is it started, How is it evolving over time?). In *Logistics 4.0* (pp. 1–14). CRC Press.

Dutta, P., Choi, T. M., Somani, S., & Butala, R. (2020). Blockchain technology in supply chain operations: Applications, challenges and research opportunities. *Transportation Research Part E: Logistics and Transportation Review*, 142, 102067.

Elhmoud, E. R., & Kutty, A. A. (2021). Sustainability assessment in aviation industry: A mini-review on the tools, models and methods of assessment. In *Proceedings of the 2nd African International Conference on Industrial Engineering and Operations Management*, Harare, Zimbabwe (pp. 7–10).

Gunasekaran, A., & Ngai, E. W. (2012). The future of operations management: An outlook and analysis. *International Journal of Production Economics*, 135(2), 687–701.

Halpern, N., Budd, T., Suau-Sanchez, P., Brathen, S., & Mwesiumo, D. (2019). Towards Airport 4.0: Airport digital maturity and transformation. In *23rd Air Transport Research Society World Conference*.

Heizer, J., Render, B., & Munson, C. (2017). *Operations Management: Sustainability and Supply Chain Management* (12th ed.). Pearson.

Koroniotis, N., Moustafa, N., Schiliro, F., Gauravaram, P., & Janicke, H. (2020). A holistic review of cybersecurity and reliability perspectives in smart airports. *IEEE Access*, 8, 209802–209834.

Lufthansa Group. (2022). *Responsibility*. Lufthansa Group. Available at: https://www.lufthansagroup.com/en/responsibility.html

Lufthansa Group Sustainability 2022 Fact Sheet. (2022). Lufthansa Group. Available at: https://www.lufthansagroup.com/en/responsibility.html

Lykou, G., Anagnostopoulou, A., & Gritzalis, D. (2018). Smart airport cybersecurity: Threat mitigation and cyber resilience controls. *Sensors*, 19(1), 19.

Michalski, K., Jurgilewicz, M., Kubiak, M., & Grądzka, A. (2020). The Implementation of selective passenger screening systems based on data analysis and behavioral profiling in the Smart Aviation Security Management – Conditions, consequences and controversies. *Journal of Security & Sustainability Issues*, 9(4), 1145–1156.

Price, R., Wrigley, C., Dreiling, A., & Bucolo, S. (2013). Design led innovation: Shifting from smart follower to digital strategy leader in the Australian airport sector. In *2013 IEEE Tsinghua International Design Management Symposium* (pp. 251–258). IEEE.

Rajapaksha, A., & Jayasuriya, N. (2020). Smart airport: A review on future of the airport operation. *Global Journal of Management and Business Research*, 20(3), 25–34.

Tan, J. H., & Masood, T. (2021). Adoption of Industry 4.0 technologies in airports – A systematic literature review. arXiv preprint arXiv:2112.14333.

Turkish Airlines – 2021 Sustainability Report (2021). www.turkishairlines.com. Retrieved August 12, 2023, from https://investor.turkishairlines.com/en/corporate-governance/sustainability

Turkish Airlines – Investor Relations. (2023). www.turkishairlines.com. Retrieved August 12, 2023, from https://investor.turkishairlines.com/en/corporate-governance/sustainability

Zaharia, S. E., & Pietreanu, C. V. (2018). Challenges in airport digital transformation. *Transportation Research Procedia*, 35, 90–99.

2 Productivity and Strategy
A Study on Airports

Mehmet Hakan Özdemir
Turkish-German University, Istanbul, Turkey

2.1 INTRODUCTION

The rapid growth of air transportation has been observed since the conclusion of World War II (Battal & Mühim, 2016). Particularly since the 1960s, there has been significant growth in global tourism. Following this rise in tourism-related demand, international air transportation saw developments that were unique to this sector of the transportation industry (Arıkan, 1998). The ongoing process of cultural, social, and economic globalization also contributes to aviation's growing significance (Carlucci et al., 2018). Hence, the functioning of air transportation plays a pivotal role in fostering economic expansion and bolstering competitiveness, thereby augmenting the general standard of living within a designated geographic region. On the contrary, inadequate airport infrastructure or the complete lack thereof within a given area might impede economic advancement (Stichhauerova & Pelloneova, 2019). Besides, airports can also be viewed as a country's entry point to the rest of the world. As a result, the image of a nation will greatly benefit when its airports are operating efficiently (Elgün et al., 2013).

Airports can be thought of as production units that use a variety of inputs to create a variety of outputs (Oum et al., 2003). For countries, the efficiency of high-cost airport operations is crucial since efficient airport operations are necessary for airports to provide a profitable business. Because of this, it is crucial to measure operational performance at specified periods to assess the efficiency of airports, and new operating strategies should be devised in light of the results. As previously stated, a wide range of inputs and outputs must be considered while evaluating the efficiency of airports (Kıyıldı & Karaşahin, 2006). It should also be emphasized that studies into airport efficiency evaluation have become increasingly popular in recent years (Olariaga & Moreno, 2019).

To examine the operational performance or productivity of an operating entity, researchers frequently employ the relative concept of efficiency. Efficiency is often defined as the capacity to use few resources and/or to produce a large volume of services in the context of production units engaged in service-related activities (Parkan, 2003; Parkan & Wu, 1999).

This study aims to evaluate the efficiency of 20 airports located in Turkey for the year 2021. The evaluation will be conducted using OCRA (operational competitiveness rating analysis), focusing specifically on airports with available data and terminals larger than 20,000 m². A literature review follows this introduction. The methodology is then presented, and the case study comes after that. The study ends with a conclusion.

2.2 LITERATURE REVIEW

In the literature, airport performance has attracted much attention. This section begins by providing a number of studies that evaluate the efficiency of Turkish airports. A study comparing Turkish airports and Spanish airports is also included. Later, studies from other countries are mentioned.

Kıyıldı and Karaşahin (2006) employed the Data Envelopment Analysis (DEA) to assess the civil aviation transportation services in 32 Turkish airports. By choosing two different output-oriented models, each airport's performance was compared to that of the others, and its efficiency score was evaluated. The inputs of the models encompass various factors, such as the number of check-in counters, quantity of X-rays, terminal area, runway area, apron area, apron area aircraft capacity, taxiway aircraft capacity, quantity of conveyor belts, and public parking capacity. The output of one model consists of the number of passengers, and the output of the other model consists of the number of aircraft.

Ulutas and Ulutas (2009) employed DEA models to evaluate the efficiency of 31 airports in Turkey in the years 2004–2005. With the use of the Analytic Network Process (ANP), the inputs used in the DEA model were predetermined. These are costs, personnel number, terminal area, passenger capacity, and airplane capacity. The outputs used in the DEA models are revenue, passenger movements, general aviation movements, commercial aviation movements, and the amount of cargo shipped.

By using data from the year 2007, Peker and Baki (2009) divided 37 airports in Turkey into two groups according to their annual passenger numbers. For each group, they analyzed the efficiency of airports using the DEA. To address this matter, the inputs taken into account were parking garage capacity, quantity of runways, sizes of airports, and personnel counts. The output considered was the number of passengers and the value of the cargo.

Koçak (2011) used the DEA model to analyze 40 Turkish airports in order to assess their efficiency for the year 2008. Inputs for this purpose were operation costs, personnel number, flight traffic, and passenger counts. Outputs included passenger counts per region, flight traffic per runway, total load, and operation costs.

Ömürbek et al. (2013) used the DEA to analyze airport efficiency, and 40 airports have been included in the study. The airports were subjected to analysis by categorizing them into three distinct groups based on their flight traffic and passenger traffic: big, medium-sized, and tiny airports. The inputs encompass various factors such as operating costs, passenger terminal area, airport passenger capacity, airport public parking capacity, apron aircraft capacity, annual aircraft capacity, total number of airport service vehicles and general-purpose vehicles, number of IT equipment, personnel count, and number of emergency equipment. On the other hand, the outputs consist of flight traffic, commercial flight traffic, cargo traffic, passenger count, and sales revenue.

Ülkü (2015) made a comparative efficiency analysis for Turkish and Spanish airports for the period 2009–2011 by using the DEA. The model incorporates various inputs, including the total runway area, personnel costs, and other operating costs. Additionally, it considers several outputs, such as the number of passengers, air traffic movement, the volume of cargo transported, and the commercial revenue generated. The study further identified the sources of inefficiencies brought on by different management strategies and other external factors based on the efficiency scores. The results indicate that, on average, Spanish airports exhibit higher levels of efficiency, whereas private involvement in operations enhances efficiency at Turkish airports.

Bolat et al. (2016) employed the DEA to determine the efficiency of 41 airports in Turkey. The inputs include quantity of check-in counters, quantity of luggage conveyors, quantity of passenger boarding gates, quantity of airways, terminal area, personnel count, and airport parking capacity, and the outputs include total count of passengers, total amount of freight transported, and total count of commercial flights. Following the DEA, an Artificial Neural Network (ANN) model was created with the aid of the same data, allowing for the prediction of the efficiency of both new and existing airports.

In their study, Örkcü et al. (2016) employed the Malmquist Productivity Index, both in its classical form and through bootstrapping techniques, to assess the operational efficiency of 21 airports located in Turkey throughout the period spanning from 2009 to 2014. The results showed that the majority of Turkish airports had an increase in efficiency and productivity over the course of the inquiry. The efficiency did, however, significantly fall between 2011 and 2012. The fundamental reason for the stagnation seen can be attributed to the substantial expansion of the physical infrastructure of Turkish airports in 2011. In 2012, there was a decrease due to the physical capacity for passenger and cargo traffic not being reflected. Despite dropping between 2011 and 2012, Turkish airport efficiency scores have risen since 2013. Additionally, the decomposition of the Malmquist index revealed that although most Turkish airports lost efficiency, they made gains in terms of technology. Three output variables and three input variables were obtained based on the literature review and the available data. The inputs of the study encompass the quantity of runways, the dimensions of runway units, and the passenger terminal area. The analysis incorporated outputs such as the yearly count of flights, the yearly volume of passengers, and the yearly volume of cargo.

In their study, Koç and Çalipinar (2017) employed the FarePrimont and Malmquist indices to determine the efficiency of 38 airports in Turkey using data spanning from 2011 to 2014. The inputs encompass the personnel number and costs, whereas the outputs encompass the passenger count and the amount of freight transported.

Şahin (2019) studied the efficiency of 42 airports under the management of the General Directorate of State Airports Authority (DHMI in Turkish acronym) in Turkey between 2014 and 2018, using the DEA and the Malmquist Total Factor Productivity Indices. The DEA has been used to establish the airports' efficiency scores for a period of five years, commencing in 2014 and ending in 2018, as the study's initial step. In the second stage, the evolution of the airports' total factor productivity values across the years was determined during the analysis period using Malmquist Total Factor Productivity Indices. By using an input-oriented model under the assumption of constant returns to scale, DEA has been set up. Five inputs were selected for the analysis: number of personnel, operating costs, terminal area, quantity of runways, and quantity of aprons. The analysis incorporates four outputs: passenger traffic, operating income, aircraft traffic, and freight traffic.

In their study, Köleoğlu and Demirel (2019) employed the DEA to assess the efficiency of airports situated in key Turkish cities that hold significance for tourism in the year 2015. The inputs for this analysis are the personnel count at airports, personnel expenses, flight count, and terminal passenger capacity. The outputs considered in this study were the number of tourists and tourism revenue. The efficiency of the airports was assessed using the input-oriented BCC (Banker, Charnes, and Cooper) model. Based on the findings of the study, it was determined that Istanbul Atatürk, Muğla Dalaman, Muğla Milas-Bodrum, Isparta, Nevşehir Kapadokya, and Denizli airports exhibited prominent efficiency amongst the 16 airports examined.

Uludağ (2020) conducted a study utilizing a hybrid methodology known as Weight Restricted EATWOS (Efficiency Analysis Technique With Output Satisficing) without consideration of satisficing levels. The study's goal was to evaluate the productivity of Turkish airports run by the DHMI between 2014 and 2018. The primary aim of the analysis was to provide recommendations for enhancing the performance of these airports. In order to conduct a comparative analysis between the outcomes derived from the suggested model and those achieved through conventional approaches, the productivity of airports was assessed using two methodologies: equally weighted EATWOS without consideration for satisficing levels, and an input-oriented DEA model assuming constant returns to scale. The inputs for this study included the total costs, quantity of runways, total passenger, cargo, and general aviation terminal areas, as well as the number of workers. The outputs considered were the passenger count and cargo traffic.

In their study, Durmuşçelebi and Kiracı (2022) employed a performance analysis to evaluate the ten most prominent airports in Turkey, focusing on their passenger volumes. The criteria weights were derived using the CRITIC (CRiteria Importance Through Intercriteria Correlation) method, followed by their utilization in the EDAS (Evaluation based on Distance from Average Solution) method by the researchers. The criteria used in the study are terminal area, distance to the city centre, quantity of runways, apron capacity, quantity of taxiways, aircraft traffic, passenger count, freight traffic, cargo traffic, revenue, costs, population, city-based gross domestic product, number of tourists, and number of staff.

Kaya et al. (2022) performed an integrated analysis that included both an eigenvector centrality analysis and a super-efficient DEA. In order to pinpoint the crucial variables affecting this efficiency, the Simar-Wilson bootstrapping method was also applied. The 39 international airports in Turkey were used as a case study to examine their efficiency and make strategic suggestions. According to the findings of the highly efficient DEA analysis and eigenvector centrality study, the Istanbul Atatürk, Adana, and Diyarbakır airports emerge as notable examples of excellence. These airports set an example for other airports with regard to operational efficiency. Additionally, Simar-Wilson bootstrapping approach-based hypothesis testing demonstrates that regional differences exist in airport efficiency. The efficiency of airports located in the Mediterranean region, which serves as a benchmark, was observed to be much higher compared to airports in the Marmara, Central Anatolia, Eastern Anatolia, and Southeast Anatolia regions.

The study conducted by Köçken et al. (2022) sought to assess the performance rankings of airports in Turkey both prior to and during the COVID-19 pandemic, which had a significant worldwide influence. The comparison is based on the relative efficiency values derived through the implementation of a three-stage DEA. This analysis utilizes data from a total of 46 airports in the year 2018 and 51 airports in the year 2020. Notably, the dataset includes Istanbul Airport, which was recently inaugurated. In the subsequent phase of the analysis, the Stochastic Frontier Analysis is employed to incorporate environmental elements such as provincial gross domestic product (GDP), tertiary sector ratio, and the number of tourists, which distinguishes it from the standard DEA. Therefore, this approach guarantees the segregation of environmental impact and chance factor from efficiency values, ensuring that business performance values alone reflect management inefficiency. The inputs for the study included the size of the apron, the number of aircraft stands, and the overall area of the passenger terminal. The output includes passenger traffic, commercial aircraft traffic, and freight traffic. An analysis was undertaken utilizing the available data to ascertain the ranking of airports according to their efficiency values, followed by subsequent comparisons.

Up to this point, the present part has encompassed studies pertaining to airports located within the country of Turkey. Following this juncture, several studies undertaken outside of Turkey will be examined.

In their study, Gillen and Lall (1997) employed the DEA to assess airport performance by developing performance indices. These indices were constructed by considering several inputs and outputs associated with airports. The main focus of this study is to establish performance indicators for airside operations, considering the airport area, quantity of runways, runway area, and personnel number as inputs and air carrier movements and commuter movements as outputs. Additionally, the study also examines terminal operations, taking into account inputs such as the number of runways, gates, terminal area, employees, baggage collection belts, and public parking places. The outputs in this case are measured by the number of passengers and the weight of cargo transported. This study investigates the performance of a panel consisting of 21 airports in the United States over a span of five years. The study used a second stage Tobit regression analysis to ascertain the variables that managers can potentially change and the magnitude of their impact on performance. According to the study, DEA is a useful technique for evaluating how well transportation infrastructure management, particularly airport management, is performing. The paucity of government investment money and the requirement for airports to become more efficient and self-sustaining are the driving forces behind the desire to privatize or commercialize airports. Airports are viewed as established businesses that ought to be allowed to run independently of the government. The effect of financial arrangements on airport efficiency, such as residual financing, is also covered in the article.

It implies that residual financing airports are more efficient than hub airline locations and gate additions, which have less of an effect on airside efficiency.

The study conducted by Wilbert et al. (2017) utilized the DEA to evaluate the operational and financial efficiency of 63 public airports that are under the management of the Brazilian Airport Infrastructure Company. According to the operational analysis, in 2010 and 2013, only 12.7% and 11.1%, respectively, of respondent terminals were efficient. Efficiency was measured based on total costs as inputs, and outputs included domestic and international passenger handling as well as domestic and international air cargo handling and mail. According to the financial analysis, 9.5% of airports were efficient in 2010, and 14.3% in 2013. This assessment was based on the consideration of total costs as inputs and revenue from both unregulated and regulated operations as outputs. The results indicate that the industry should prioritize efforts to address bottlenecks and enhance productivity by implementing modernization measures, increasing investment in infrastructure, and achieving a balance between income and expenses.

In their study, Carlucci et al. (2018) conducted an analysis of the efficiency of 34 Italian airports over a period of ten years, from 2006 to 2016. The authors employed the DEA to evaluate the airports' overall efficiency, as well as their pure technical efficiency and scale efficiency. Labour costs, invested capital, and other expenses were used as inputs, and passenger movements, cargo, aircraft movements, revenues from aeronautical activities, revenues from handling activities, and revenues from commercial activities were used as outputs. The goal was to determine how various factors affect the efficiency and economic sustainability of regional airports. The findings show that the technical and scale efficiency of Italian airports are significantly impacted by the size of the airport, the presence of low-cost carriers, and cargo traffic. Stated differently, the privatization and deregulation of air transport have the potential to enhance the sustainability and efficiency of regional airports.

In order to evaluate airport performance and efficiency using MACBETH (Measuring Attractiveness by a Categorical Based Evaluation Technique), Baltazar et al. (2018) suggested a global decision support (GDS) model. With criteria weights and value scales drawn from expert opinions and comparisons of various reference levels and performance profiles, the model entails building a hierarchical additive value model. To detect shortcomings that require immediate intervention and remedial measures for continual improvement, a management system can be connected with the outputs of the GDS model. The study highlights the value of airport benchmarking for a range of stakeholders, including business, operational management, airlines, passengers, and government agencies. The GDS approach permits self-benchmarking for a predetermined amount of time or peer benchmarking among a group of direct competitors. The paper's overall conclusion emphasizes the value of evaluating and enhancing airport performance and efficiency through the use of an extensive decision support model.

Stichhauerova and Pelloneova (2019) assessed the technical efficiency of Germany's 27 major airports by using the DEA. The paper's first section reviews the literature on the use of the DEA technique for airport and air transportation performance evaluation. A comprehensive compilation of inputs and outputs employed by international authors in their works to evaluate airport performance has been curated specifically for this DEA. The actual research methodology is covered in the second section of the paper. The primary data source was the annual reports from various airports for 2016. The inputs used were the quantity of runways, the number of employees, and the airport's area. Two variables were selected to be the outputs: the quantity of cargo and the number of aircraft movements. Thirteen airports have been found to be able to efficiently convert the inputs into outputs by using input-oriented CCR (Charnes, Cooper, and Rhodes) and BCC models. This is because they implement the right procedures and best practices in their operations management. It is also possible to characterize five airports as establishments that have reached their optimal and most productive size.

2.3 METHODOLOGY

The entropy method was used in this study to establish the input and output weights, and OCRA was then used to calculate the efficiency ranking. A thorough explanation of these steps is provided below.

2.3.1 ENTROPY METHOD

Assuming that there are m alternatives and n criteria, the entropy method uses the following equations to calculate the weight of each criterion (Alinezhad & Khalili, 2019; Huang, 2008).

1) Eq. (2.1) is used to build the decision matrix $\mathbf{X}$.

$$\mathbf{X} = \begin{bmatrix} x_{11} & \cdots & x_{1j} & \cdots & x_{1n} \\ \vdots & \ddots & \vdots & \ddots & \vdots \\ x_{i1} & \cdots & x_{ij} & \cdots & x_{in} \\ \vdots & \ddots & \vdots & \ddots & \vdots \\ x_{m1} & \cdots & x_{mj} & \cdots & x_{mn} \end{bmatrix}; \ i = 1, 2, \ldots, m \ \ j = 1, 2, \ldots, n \tag{2.1}$$

2) Eq. (2.2) is used to normalize the decision matrix.

$$p_{ij} = \frac{x_{ij}}{\sum_{i=1}^{m} x_{ij}}; \ j = 1, 2, \ldots, n \tag{2.2}$$

3) Eq. (2.3) is used to compute the degree of entropy for each criterion, denoted by e_j.

$$e_j = -\frac{1}{\ln(m)} \sum_{i=1}^{m} p_{ij} \cdot \ln(p_{ij}); \ j = 1, 2, \ldots, n; \ 0 \le e_j \le 1 \tag{2.3}$$

4) Eq. (2.4) is used to calculate the variation rate of the degree of entropy for each criterion, denoted by d_j.

$$d_j = 1 - e_j, j = 1, 2, \ldots, n \tag{2.4}$$

5) Lastly, Eq. (2.5) is employed to calculate the weights w_j.

$$w_j = \frac{d_j}{\sum_{j=1}^{n} d_j} \tag{2.5}$$

2.3.2 OCRA

OCRA was introduced by Parkan (1994). According to Peters and Zelewski (2010), it is used to address efficiency analysis and performance measurement problems. The scaled performance indices using OCRA are computed with the following steps (Işık & Adalı, 2016; Kundakcı, 2019; Ozcalici & Bumin, 2020; Peters & Zelewski, 2010).

1) Eq. (2.6) is used to build the decision matrix **X**,

$$\mathbf{X} = \begin{bmatrix} x_{11} & \cdots & x_{1j} & \cdots & x_{1n} \\ \vdots & \ddots & \vdots & \ddots & \vdots \\ x_{i1} & \cdots & x_{ij} & \cdots & x_{in} \\ \vdots & \ddots & \vdots & \ddots & \vdots \\ x_{m1} & \cdots & x_{mj} & \cdots & x_{mn} \end{bmatrix}; i = 1, 2, \ldots, m \ \ j = 1, 2, \ldots, n \tag{2.6}$$

where m denotes the quantity of decision-making units (DMU) or alternatives and n indicates the total count of inputs (non-beneficial criteria) and outputs (beneficial criteria).

2) The calculation of the unscaled input index $\bar{I}_i$ for each DMU is performed using Eq. (2.7),

$$\bar{I}_i = \sum_{j=1}^{g} w_j \frac{\max_i (x_{ij}) - x_{ij}}{\min_i (x_{ij})}; i = 1, 2, \ldots, m \tag{2.7}$$

where g denotes the quantity of inputs and w_j indicates the weight assigned to the jth input.

3) Employing Eq. (2.8), the scaled input index $\bar{\bar{I}}_i$ is calculated for each DMU.

$$\bar{\bar{I}}_i = \bar{I}_i - \min_i (\bar{I}_i); i = 1, 2, \ldots, m \tag{2.8}$$

4) Using Eq. (2.9), the unscaled output index $\bar{O}_i$ is calculated for each DMU,

$$\bar{O}_i = \sum_{j=g+1}^{n} w_j \frac{x_{ij} - \min_i\left(x_{ij}\right)}{\min_i\left(x_{ij}\right)}; i = 1, 2, \ldots, m \qquad (2.9)$$

with $(n - g)$ representing the quantity of outputs and w_j representing the weight of the jth output.

5) Each DMU's scaled output index $\bar{\bar{O}}_i$ is calculated using Eq. (2.10).

$$\bar{\bar{O}}_i = \bar{O}_i - \min_i\left(\bar{O}_i\right); i = 1, 2, \ldots, m \qquad (2.10)$$

6) Each DMU's scaled performance index P_i is calculated employing Eq. (2.11).

$$P_i = \left(\bar{\bar{I}}_i + \bar{\bar{O}}_i\right) - \min_i\left(\bar{\bar{I}}_i + \bar{\bar{O}}_i\right); i = 1, 2, \ldots, m \qquad (2.11)$$

7) The ranking of the efficiency of the DMUs is determined based on the scaled performance index. The DMU that exhibits the greatest scaled performance index is considered to be the most efficient.

2.4 CASE STUDY: EFFICIENCY OF TURKISH AIRPORTS

The present study aims to evaluate the efficiency of 20 airports located in Turkey for the year 2021. The assessment will be carried out utilizing OCRA, with a specific emphasis on airports that possess accessible data and terminals that surpass an area of 20,000 m². The literature review section was used to determine the four inputs and four outputs employed in the efficiency analysis. Table 2.1 shows these inputs and outputs with their abbreviations.

TABLE 2.1
Inputs and Outputs Used in OCRA

Input 1 (I1)	Number of runways
Input 2 (I2)	Terminal area (m²)
Input 3 (I3)	Total number of staff
Input 4 (I4)	Total costs (TL in thousands)
Output 1 (O1)	Total number of flights
Output 2 (O2)	Total number of passengers
Output 3 (O3)	Total freight transport (tons)
Output 4 (O4)	Service sales revenue (TL in thousands)

TABLE 2.2
IATA Codes of the Airports

Nr.	IATA Code	Airport
1	IST	Istanbul
2	ADB	Izmir Adnan Menderes
3	AYT	Antalya
4	DLM	Muğla Dalaman
5	BJV	Muğla Milas-Bodrum
6	DIY	Diyarbakır
7	GZT	Gaziantep
8	HTY	Hatay
9	KSY	Kars Harakani
10	MQM	Mardin
11	ERC	Erzincan Yıldırım Akbulut
12	KYA	Konya
13	TZX	Trabzon
14	AJI	Ağrı Ahmed-i Hani
15	EDO	Balıkesir Koca Seyit
16	ADF	Adıyaman
17	KCM	Kahramanmaraş
18	BAL	Batman
19	OGU	Ordu-Giresun
20	VAS	Sivas Nuri Demirağ

Table 2.2 presents the International Air Transport Association (IATA) codes corresponding to the airports that have been included in the analysis, and Table 2.3 shows the data of the airports obtained from DHMI's 2021 annual report (DHMI, 2021).

TABLE 2.3
Data of the Airports

IATA Codes	I1	I2	I3	I4	O1	O2	O3	O4
IST	5	1,370,000	264	118,031	279,509	37,181,907	1,529,778	1,380,745
ADB	2	310,978	650	493,362	58,299	7,569,054	109,133	578,546
AYT	3	178,637	631	878,216	140,576	22,013,861	256,948	2,321,573
DLM	1	244,406	360	207,580	31,722	2,323,874	23,941	380,789
BJV	2	110,613	298	891,538	29,755	2,909,337	30,639	473,666
DIY	1	86,571	159	99,951	9,073	1,349,124	11,900	36,584
GZT	1	72,593	271	130,858	16,469	1,859,524	20,041	64,641
HTY	1	43,688	139	64,632	7,650	898,377	9,650	25,935

TABLE 2.3   (Continued)

IATA Codes	I1	I2	I3	I4	O1	O2	O3	O4
KSY	2	35,946	99	50,027	3,365	453,366	4,322	8,898
MQM	1	33,150	102	48,159	4,101	580,889	5,311	9,660
ERC	1	27,132	111	52,146	2,442	289,755	2,880	8,333
KYA	2	24,175	181	66,016	6,424	662,397	7,724	25,734
TZX	1	23,745	265	117,437	20,958	2,642,327	26,760	91,978
AJI	1	23,676	81	43,264	1,895	250,600	2,342	5,651
EDO	1	23,240	160	56,231	20,751	265,555	2,241	7,378
ADF	1	23,011	107	51,821	1,341	132,535	1,282	4,132
KCM	1	22,330	141	52,319	2,230	171,785	1,535	5,258
BAL	1	20,741	99	40,114	3,655	526,546	4,994	8,847
OGU	1	20,250	134	59,939	5,810	740,653	6,411	18,437
VAS	1	20,047	134	50,353	4,118	410,437	3,956	8,765

TABLE 2.4
Weights of the Inputs and Outputs

I1	I2	I3	I4	O1	O2	O3	O4
0.02	0.14	0.027	0.086	0.14	0.162	0.252	0.173

Table 2.4 displays the weights that were determined for the inputs and outputs using the entropy method.

Regarding the inputs, Table 2.4 indicates that the terminal area (I2) has the highest weight. This is followed by total costs (I4), the total number of staff (I3), and the number of runways (I1). As for the outputs, it can be observed that the total freight

TABLE 2.5
Efficiency Ranking of the Airports

IATA Code	Airport	Scaled Performance Index	Ranking
IST	Istanbul	422.14	1
ADB	Izmir Adnan Menderes	57.02	3
AYT	Antalya	185.13	2
DLM	Muğla Dalaman	24.06	5
BJV	Muğla Milas-Bodrum	29.23	4
DIY	Diyarbakır	5.17	8
GZT	Gaziantep	9.33	7
HTY	Hatay	3.96	9

(Continued)

TABLE 2.5   (Continued)

IATA Code	Airport	Scaled Performance Index	Ranking
KSY	Kars Harakani	1.29	16
MQM	Mardin	1.80	13
ERC	Erzincan Yıldırım Akbulut	0.77	17
KYA	Konya	3.26	10
TZX	Trabzon	13.58	6
AJI	Ağrı Ahmed-i Hani	0.50	18
EDO	Balıkesir Koca Seyit	2.49	12
ADF	Adıyaman	0	20
KCM	Kahramanmaraş	0.23	19
BAL	Batman	1.69	14
OGU	Ordu-Giresun	2.81	11
VAS	Sivas Nuri Demirağ	1.36	15

transport (O3) has the highest weight. Next in order are service sales revenue (O4), the total number of passengers (O2), and the total number of flights (O1).

Table 2.5 displays the scaled performance indices and OCRA ranking of the airports, utilizing the input and output weights as shown in Table 2.4.

According to the results presented in Table 2.5, Istanbul Airport holds the top position. Upon reviewing the data shown in Table 2.3, it becomes evident that this result can be attributed to this airport's far greater outputs, despite the fact that the inputs of I1 and I2 surpass those of the other airports. Despite the high inputs of the first five airports in Table 2.3, their high outputs position them at the forefront of the ranking. In contrast, Adıyaman Airport ranks the lowest. Based on the data presented in Table 2.3, it can be observed that airports exhibiting lower values in certain inputs tend to yield higher outputs, thereby resulting in a lower ranking for Adıyaman airport. Airports Kahramanmaraş and Ağrı Ahmed-i Hani rank 19th and 18th, respectively, due to their extremely low output values relative to their inputs.

2.5 CONCLUSION

Airports are often considered to be a country's primary point of contact with the rest of the world. Consequently, a country's reputation will be significantly improved whenever its airports function in an efficient manner. The efficiency of high-cost airport operations is of utmost importance, given that efficient airport operations are essential for airports to be able to deliver lucrative business. As a result of this, it is essential to evaluate the efficiency of airports by measuring operational performance at predetermined intervals, and new operating strategies should be developed according to the results.

This study aimed to evaluate the efficiency of 20 airports in Turkey for the year 2021. The evaluation was conducted using OCRA, which incorporated four inputs and four outputs. The determination of input and output weights was conducted using the entropy method. Upon closer examination of the literature review, it becomes evident that the efficiency measurement of airports is largely conducted using the DEA. This study employed the application of OCRA with the objective of making a scholarly contribution to the existing literature. For further research, it is recommended to employ other weighting techniques and efficiency assessment approaches, which can afterwards be utilized to compare the resultant rankings with those reported in the current study.

REFERENCES

Alinezhad, A., & Khalili, J. (2019). *New methods and applications in multiple attribute decision making (MADM)* (Vol. 277). Springer.

Arıkan, I. (1998). Havayolu Ulaşımı ile Turizm İlişkisi ve Havaalanları. *Anatolia: Turizm Araştırmaları Dergisi, 9*(2), 46–54.

Baltazar, M. E., Rosa, T., & Silva, J. (2018). Global decision support for airport performance and efficiency assessment. *Journal of Air Transport Management, 71*, 220–242.

Battal, Ü., & Mühim, S. A. (2016). Havayolu Taşımacılığında Yakıt Anlaşmalarında Riskten Korunma Yöntemleri ve Türkiye Uygulaması. *Finans Politik ve Ekonomi Yorumlar*, (611), 39–56.

Bolat, B., Temur, G. T., & Gürler, H. (2016). Türkiye'deki Havalimanlarının Etkinlik Tahmini: Veri Zarflama Analizi ve Yapay Sınır Ağlarının Birlikte Kullanımı. *Ege academic Review, 16*.

Carlucci, F., Cirà, A., & Coccorese, P. (2018). Measuring and explaining airport efficiency and sustainability: Evidence from Italy. *Sustainability, 10*(2), 400.

DHMI (2021). Devlet Hava Meydanları İşletmesi DHMI 2021 Faaliyet Raporu, https://www.dhmi.gov.tr/Lists/FaaliyetRaporlari/Attachments/22/Faaliyet%20Raporu-31.05.2021web%20(3).pdf

Durmuşçelebi, C., & Kiracı, K. (2022). Türkiye'de Havaalanı Performansının CRITIC temelli EDAS Yöntemiyle Analizi. *Anemon Muş Alparslan Üniversitesi Sosyal Bilimler Dergisi, 10*(2), 837–856.

Elgün, A., Babacan, E., Kozak, M., & Babat, D. (2013). Yeni tüketim mekânları olarak havalimanı terminalleri. *Anatolia: Turizm Araştırmaları Dergisi, 24*(1), 70–82.

Gillen, D., & Lall, A. (1997). Developing measures of airport productivity and performance: an application of data envelopment analysis. *Transportation Research Part E: Logistics and Transportation Review, 33*(4), 261–273.

Huang, J. (2008). Combining entropy weight and TOPSIS method for information system selection. In *2008 IEEE Conference on Cybernetics and Intelligent Systems*, 1281–1284. IEEE.

Işık, A. T., & Adalı, E. A. (2016). A new integrated decision making approach based on SWARA and OCRA methods for the hotel selection problem. *International Journal of Advanced Operations Management, 8*(2), 140–151.

Kaya, G., Aydın, U., Karadayı, M. A., Ülengin, F., Ülengin, B., & İçken, A. (2022). Integrated methodology for evaluating the efficiency of airports: A case study in Turkey. *Transport Policy, 127*, 31–47.

Kıyıldı, R., & Karaşahin, M. (2006). Türkiye'deki hava alanlarının veri zarflama analizi ile altyapı performansının değerlendirilmesi. *Süleyman Demirel Üniversitesi Fen Bilimleri Enstitüsü Dergisi, 10*(3), 391–397.

Koç, E., & Çalipinar, H. (2017). Fareprimont Ve Malmquist Verimlilik Endeksleri İle Türk Havalimanlarının Etkinliklerinin Karşılaştırılması. *Journal of Academic Value Studies, 3*(8), 77–87.

Koçak, H. (2011). Efficiency examination of Turkish airports with DEA approach. *International Business Research, 4*(2), 204.

Köçken, K., Timor, M., & Karakaplan, M. U. (2022). COVID-19 Pandemisi Öncesinde ve Pandemi Döneminde Türkiye'deki Havalimanı Etkinliklerinin Üç Aşamalı Veri Zarflama Analizi İle Belirlenmesi. *Avrupa Bilim ve Teknoloji Dergisi*, (35), 643–652.

Köleoğlu, N., & Demirel, E. (2019). Türkiye'nin önemli turizm kentlerindeki havalimanlarının etkinliklerinin veri zarflama analizi yöntemiyle ölçülmesi. *Seyahat ve Otel İşletmeciliği Dergisi*, *16*(3), 352–365.

Kundakcı, N. (2019). A comparative analyze based on EATWOS and OCRA methods for supplier evaluation. *Alphanumeric Journal*, *7*(1), 103–112.

Olariaga, O. D., & Moreno, L. P. (2019). Measurement of airport efficiency. The case of Colombia. *Transport and Telecommunication*, *20*(1), 40–51.

Ömürbek, N., Demirgubuz, M. Ö., & Tunca, M. Z. (2013). Hizmet Sektöründe Performans Ölçümünde Veri Zarflama Analizinin Kullanımı: Havalimanları Üzerine Bir Uygulama. *Süleyman Demirel Üniversitesi Vizyoner Dergisi*, *4*(9), 21–43.

Örkcü, H. H., Balıkçı, C., Dogan, M. I., & Genç, A. (2016). An evaluation of the operational efficiency of Turkish airports using data envelopment analysis and the Malmquist productivity index: 2009–2014 case. *Transport Policy*, *48*, 92–104.

Oum, T. H., Yu, C., & Fu, X. (2003). A comparative analysis of productivity performance of the world's major airports: Summary report of the ATRS global airport benchmarking research report—2002. *Journal of Air Transport Management*, *9*(5), 285–297.

Ozcalici, M., & Bumin, M. (2020). An integrated multi-criteria decision making model with Self-Organizing Maps for the assessment of the performance of publicly traded banks in Borsa Istanbul. *Applied Soft Computing*, *90*, 106166.

Parkan, C. (1994). Operational competitiveness ratings of production units. *Managerial and Decision Economics*, *15*(3), 201–221.

Parkan, C. (2003). Measuring the effect of a new point of sale system on the performance of drugstore operations. *Computers & Operations Research*, *30*(5), 729–744.

Parkan, C., & Wu, M. L. (1999). Measuring the performance of operations of Hong Kong's manufacturing industries. *European Journal of Operational Research*, *118*(2), 235–258.

Peker, I., & Baki, B. (2009). Veri zarflama analizi ile Türkiye havalimanlarında bir etkinlik ölçümü uygulaması. *Çukurova Üniversitesi Sosyal Bilimler Enstitüsü Dergisi*, *18*(2), 72–88.

Peters, M. L., & Zelewski, S. (2010). Performance measurement mithilfe des Operational Competitiveness Ratings (OCRA). *WiSt-Wirtschaftswissenschaftliches Studium*, *39*(5), 224–229.

Şahin, I. E. (2019). Türkiye'deki Havalimanlarının Veri Zarflama Analizi ve Malmquist Toplam Faktör Verimliliği Endeksleri İle Finansal Etkinliklerinin Analizi. *Selçuk Üniversitesi Sosyal Bilimler Enstitüsü Dergisi*, (42), 33–47.

Stichhauerova, E., & Pelloneova, N. (2019). An efficiency assessment of selected German airports using the DEA model. *Journal of Competitiveness*, *11*(1), 135–151.

Ülkü, T. (2015). A comparative efficiency analysis of Spanish and Turkish airports. *Journal of Air Transport Management*, *46*, 56–68.

Uludağ, A. S. (2020). Measuring the productivity of selected airports in Turkey. *Transportation Research Part E: Logistics and Transportation Review*, *141*, 102020.

Ulutas, B., & Ulutas, B. (2009, August). An analytic network process combined data envelopment analysis methodology to evaluate the performance of airports in Turkey. In *10th International Symposium on the Analytic Hierarchy/Network Process*, July (Vol. 29).

Wilbert, M. D., Serrano, A. L. M., Flores, M. R., Damasceno, R., & Franco, V. R. (2017). Efficiency analysis of airports administered by infraero from 2003 to 2013. *Applied Mathematical Sciences*, *11*(25), 1221–1238.

3 Air Passenger Demand Forecasting Using Machine Learning Algorithms
A Case Study of Turkey in Airline Industry

Beyzanur Cayir Ervural
Necmettin Erbakan University, Konya, Turkey

3.1 INTRODUCTION

The aviation sector occupies a strategic position in the global competitive market conditions due to its fragile dynamics, which directly concerns various fields such as the transportation management, cargo logistics, financial planning, and import/export activities [1]. The air transport model offers a reliable and fast option for conveying passengers and supplies over long distances, despite disadvantages such as capacity, operating costs, high-level of technical support and information infrastructure, and high airfares [2]. In order to smoothly manage all multidimensional factors, aviation authorities should coordinate airflow traffic using demand forecast projections. Well-structured demand forecasting models provide guidance for future situations [3].

Parallel to the growth in the global economy and mobility, the volume of passengers travelling by air, especially in the civil aviation sector, has increased dramatically [4]. The total revenue of the aviation industry in the last 20 years has exceeded $838 million dollars, and recorded passengers exceed 4.54 billion [5]. It was also stated that due to the pandemic, the number of flights in Europe decreased by 55% in 2020 compared to 2019, and the number of passengers reduced by 1.7 billion [6].

Air transportation has become one of the most significant and strategic advantages with the increasing importance of speed and flexibility in global platforms where today's competitive conditions intensify [7]. Accurate forecasting models assist airlines in making strategic decisions in the future, such as increasing or decreasing capacity [8]. Thus, many important issues such as purchase of additional aircraft, the expansion of additional flight crew, and the addition of a new flight route, can be clearly identified [9]. To evaluate the properties of future demand, it is essential to grow reliable forecasts of airport actions [10]. Furthermore, accurate forecasting of future air transport demand is critical in long- and medium-term airport master plans [11]. Air transportation has caught a rapid growth trend in recent years thanks to the arrangements made and the opportunities provided. Although there have been periods of decline surrounding political events, epidemics, or disasters, it has generally resumed the upward trend [12]. Demand forecasting projections and accuracy are important because air transport impacts other sectors such as transportation, tourism, and the economy [13].

To efficiently plan for the future, it is important to develop accurate forecasting models and to correctly identify the components that affect the forecast of the industry. Due to the unique dynamics of the aviation industry, it is necessary to work meticulously on the most successful model proposals. The success of the developed demand forecasting model should reflect the minimum level of deviation from the actual data to avoid any disturbance since deviations in the forecast can be very costly for airport executives [14]. Underestimating can lead to congestion, delays, and loss of customers from the resulting inadequate airport services. On the other hand, overestimating passenger demand can also create serious financial difficulties for airport authorities. Therefore, the accuracy of the prediction model is a very sensitive and critical issue [15]. For a successful and accurate forecasting model, the factors to be included in the model should be assigned correctly. One of the most essential dynamics affecting demand in air transportation is economic conditions. Other determinants related to geographical and demographic factors, social factors and other political and extraordinary issues [13].

Turkey is trying to increase its power in airline passenger transportation by investing heavily in the aviation sector. When global passenger transport traffic, Turkey shows considerable success despite all adverse conditions. According to the 2020 statistics given in Tables 3.1 and 3.2, Turkey is placed 9th in the world passenger rankings, and 3rd in the European passenger rankings [16].

It is therefore apparent how important and valuable accurate forecasting models are in the aviation industry, and thus advanced forecasting models are needed. It is difficult to predict airline passenger demand easily and precisely because it is hard to develop a suitable model due to the demonstration of complex data structure such as non-stationary data type, volatility, seasonality, non-linearity and missing data [17].

The main purpose of this study is to improve high-accuracy forecasting models to predict the future patterns of Turkey's airline passenger demand. The most popular quantitative forecasting approaches are econometric, statistical, and artificial intelligence approaches, all of which are used in a variety of applications. Specifically, apart from the traditional forecasting methods outlined in the study, we focus on developing successful forecasting models using machine learning algorithms, then on analyzing their impact on air passenger demand, and developing solutions for accurate forecasts.

The aim of this chapter is to develop various forecasting models to show the success and the quality of prediction accuracy of each model and comparison of each model by using traditional forecasting methods such as multiple linear regression (MLR) and time series (ARIMA (0,1,0)), and recently developed machine learning

TABLE 3.1
World Passenger Traffic Ranking 2020 [16]

Rank	Country	Total Passengers	2020/2019
1	USA	680,731,488	−58.8%
2	China	415,584,698	−34.2%
3	Indian	118,987,811	−60.3%
4	Russian	94,062,404	−45.8%
5	Japan	87,958,206	−65.1%
6	Brazil	83,919,464	−52.0%
7	Mexican	75,254,844	−49.8%
8	UK	73,596,256	−75.5%
9	Turkey	68,058,915	−63.0%
10	South Korea	64,898,048	−58.8%
11	France	63,116,503	−68.0%
12	Spain	58,915,435	−74.4%
13	Germany	55,190,373	−77.9%
14	Italy	50,237,057	−73.1%
15	Thailand	46,618,200	−67.4%
16	Canada	40,630,665	−72.3%
17	UAE	36,135,689	−70.6%
18	Saudi Arabia	35,353,036	319.8%
19	Australia	33,792,079	−72.3%
20	Holland	23,500,907	−70.8%

TABLE 3.2
European Passenger Traffic Ranking [16]

Rank	Country	Total Passengers	2020/2019
1	Russian	94,062,404	−45.8%
2	UK	73,596,256	−75.5%
3	Turkey	68,058,915	−63.0%
4	France	63,116,503	−68.0%
5	Spain	58,915,435	−74.4%
6	Germany	55,190,373	−77.9%
7	Italy	50,237,057	−73.1%
8	Holland	23,500,907	−70.8%
9	Norway	21,576,389	−60.2%
10	Greece	19,713,837	−69.7%

algorithms such as artificial neural networks (ANN), support vector machines (SVM), and random forest (RF) algorithms for air passenger demand forecasting in Turkey. The case of Turkey is presented by using air passenger transport data covering the 2000–2022 period. ANN and SVR were selected as supervised machine learning algorithms due to their expanding prominence and solution ability in the complicated, nonlinear, and volatile data structure. RF is a kind of multi-level decision tree and another type of supervised machine learning tool, and it gives better results than classical regression analysis.

From literature review, it can be seen that there are several studies that use various forecasting methods on air passenger demand traffic estimates, but a serious gap remains in this regard. It has been observed that insufficient studies have been conducted on air passenger demand forecasting, particularly in Turkey, and there is no comprehensive evaluation comparing it with machine learning methods and traditional approaches. It is expected to contribute to the literature in terms of the application area/country of the demand forecasting study, the data originality used and the classical and advanced methods employed, and it is aimed to eliminate the stated deficiencies.

The remainder of this study proceeds as follows. The literature review is summarized in Section 3.2. Background research methodology information is detailed in Section 3.3. Application of the forecasting methodology is discussed in Section 3.4. Obtained results are compared in Section 3.5. Finally, a short conclusion and a discussion for future remarks provided in Section 3.6.

3.2 LITERATURE REVIEW

Airline demand forecasting plays a key role not only for tactical and operational level decisions, but also as an objective evaluation methodology for long-term strategic investment decisions. There are numerous methodologies based on forecasting techniques to predict air passenger demand in the scientific literature. Although traditional approaches, artificial intelligence methods, and hybrid techniques are applied in demand forecasting studies, it is seen that time series analysis and causality models are mostly preferred in airline passenger demand forecasting tools. Table 3.3 provides a summary of current studies on of air passenger demand forecasting.

Jin et al. [18] put forward an integrated method which consists of variational mode decomposition, autoregressive moving average model, and kernel extreme learning machine for the short-term forecasting. The suggested model provides superior results in terms of accuracy and robustness. Kim and Shin [19] present a forecasting model for short-term air passenger flow utilizing big data with a causal analysis. Gelhausen et al. [20] propose a classical four-step passenger forecasting model based on co-integration theory to analyze travel behaviour at the German airport in detail. The DLR-Demand Model proceeds the stages of trip generation, trip distribution, modal split, and trip assignment to monitor the classical four-step procedure of models applied to adapt and predict traffic. Gunter and Zekan [21] utilize a GVAR model on air passenger forecasting for the top 20 airports of the world, as well as the Asia-Pacific and Latin America-Caribbean districts. In the study, some important factors are employed such as economic determinants, expense, and income.

TABLE 3.3
Literature Review

Authors	Goal	Method	Findings
Tang et al. [22]	Making weekly passenger traffic forecasts with pandemic-related variables	Machine learning model, times series model	The results showed that inclusion of pandemic variables lessen the model error by 27.7% compared to the basic model.
Tirtha et al. [33]	Assess effect of COVID-19 on airline demand	Linear mixed model	In the model different variables are utilized such as COVID-19 components, demographic and environment characteristics, spatial variables, and temporal elements and the obtained predictions were assessed under several scenarios for future COVID-19 spread
Gunter and Zekan [21]	examine air passenger forecasting for top 20 airports of the world and the Asia-Pacific and Latin America-Caribbean regions	GVAR model	To provide allowing for joint analysis of micro and macro perspectives.<XX>GVAR model succeeds most accurate predictions for few airports.
Lin Long et al. [23]	Choose the appropriate exploration inquiry that can predict arrival air passengers in Singapore Changi Airport	Neural granger causality model	The obtained results show that neural granger queries inputs to a prediction model provided higher prediction success.
Albayrak et al. [26]	Examine the factors of air passenger traffic in Turkey	Panel data estimation methodology	The obtained results show that GDP/capita, population, distance to alternative airports, tourism, leading cities, and international migrations all help with extra air traffic
Solvoll et al. [14]	Make air traffic demands in Norwegian airport	Quantitative forecasting models include 'elasticity', 'analogy', and 'transport' models	Analogies are better than elasticity methods for big variations in the system as a forecasting method
Jin et al. [18]	Forecast the air passenger demand with a novel hybrid method	Hybrid method consists of VMD-ARMA/ KELM-KELM for the short-term forecasting model	Provides better results in terms of accuracy and robustness. The new method has a more clearly superiority than other benchmark models concerning both accuracy and strength analysis.

(Continued)

TABLE 3.3   (Continued)

Authors	Goal	Method	Findings
Sun et al. [40]	Forecast air passenger flows in Beijing International Airport	Nonlinear vector auto-regression neural network approach	The proposed approach shows better performance compared to single models like times series in terms of accuracy level.
Suh and Ryerson [28]	Overcome demand uncertainty and try to predict probability of passenger volume contraction	Statistical forecasting model	Develop accuracy level cooperating previous prediction errors of airport peers.
Nieto et al. [32]	Air transportation passenger demand forecasting	Integrated method consist of ARIMA + GARCH + Bootstrap time series method	The performance of the proposed hybrid method is better than other forecasting approaches.
Gelhausen et al. [20]	Co-integration theory was applied to analyze travel behaviour at the German airport in detail	Co-integrated regression functions conventional four-step procedure of models employed for traffic forecasting	The first phase of the model which have been econometrically projected considering time series. The classical method has a robust base in cross-sectional analyses
Rajendran et al. [37]	Investigate demand of air taxi urban mobility services	Logistic regression, ANN, RF, and gradient boosting	Gradient boosting method outperforms other methods in terms of accuracy level.
Qin et al. [38]	Forecast monthly passenger flow in China using hybrid novel methods	Two novel hybrid methods combining seasonal-trend decomposition procedures and adaptive boosting approach	The proposed methods provide higher accuracy compared to traditional forecasting methods.
Vadlamani et al. [39]	Assess and forecast entry design of Southwest Airlines throughout several city pairs	Logistic regression, decision tree, support vector machine, Skope rules and random forest	Logistic regression and decision tree models well implemented in forecasting the entry behaviours compared to others.

This is the first study to provide financially interpretable air passenger demand forecasts while also letting for joint analysis of micro (i.e., forecasts for only airports) and macro viewpoints. GVAR model succeeds most accurate predictions for few airports.

Suryani et al. [10] investigated air passenger demand forecasting of a system performance using system dynamics framework in Taiwan Taoyuan International Airport based on some indicators such as ticket price impact, quality of service impact, GDP, population, number of flights per day and dwell time. All these parameters have an important impact on defining air passenger volume. Tang et al.

[22] examined the correlation between COVID-19 pandemic movements and passenger traffic at a main airport terminal in China was examined using daily statistics. And the SHapley Additive exPlanations scores was utilized to measure support of input pandemic variables. They developed a machine learning model to predict weekly passenger traffic using pandemic-related factors and other forecasters such as previous numbers of passengers, time-related inputs, and quarterly GDP. The results showed that addition of pandemic factors lessen the model error by 27.7% compared to the basic model. Li Long et al. [23] studied the employment of a neural granger causality model to select the appropriate search query that can predict arrival air passengers in Singapore Changi Airport and the obtained results show that neural granger causality can specify google trend query forecasters. Neural granger queries inputs to a prediction model provided higher prediction success.

Sun et al. [24] proposed a nonlinear vector auto-regression neural network approach to forecast air passenger flows in Beijing International Airport. According to the obtained results the proposed approach shows better performance compared to single models like times series in terms of accuracy level. Xie et al. [25] developed two integrated methods consist of seasonal decomposition and least squares support vector regression model for short-term forecasting of air passenger at Hong Kong International Airport. The suggested integrated models outperform other time series models in terms of performance criteria.

Albayrak et al. [26] researched the components of air passenger traffic in emerging country Turkey between 2004 and 2014 at provincial level using a panel data estimation approach. According to the results, the factors of Turkey's air passenger traffic are similar to those of other developed countries, and help more air traffic are GDP/capita, population, distance to other airports, tourism, top cities, and international movements. Xiao et al. [27] suggested a novel hybrid-three staged method includes singular spectrum analysis, adaptive-network-based fuzzy inference system and improved particle swarm optimization methods because of the complex and uncertain behaviour of data structures in air passenger traffic prediction. The obtained results show the proposed method performs better than competing models in terms of prediction errors. Suh and Ryerson [28] developed a statistical forecasting model to overcome demand uncertainty and try to predict probability of passenger volume contraction and to develop accuracy level cooperating previous prediction errors of airport peers.

Solvoll et al. [14] focused on different predicting methods to make air traffic demands in Norwegian airport. Analogies are better than elasticity methods for big variations in the system as a forecasting method. Carmona-Benítez et al. [29] gave an econometric dynamic model to forecast passenger demand for air transport industry in Mexican airport. The author utilized the panel data Arellano-Bover method to apply EDM because of the data structure and they validated method with Sargan test and the Arellano-Bond Autocorrelation test. Fildes et al. [30] applied several methods to forecast short- to medium-term air passenger traffic flows by means of multiple error measures. The econometric models combined autoregressive distributed lag models, time-varying parameter models, and an automatic method for econometric model. Also a vector autoregressive model was implemented for unconditional predictions. Grubb and Mason [31] examined a long lead-time projecting of air passengers in UK using Holt–Winters decomposition from 1949 to 2001. Nieto et al. [32] developed a hybrid ARIMA + GARCH + Bootstrap time series method for air transportation passenger demand predicting first time. The performance of the recommended methods is validated using other approaches.

Tirtha et al. [33] evaluated effect of COVID-19 on air company demand applying a linear mixed model between January 2019 and December 2020. In the model different variables are utilized such as COVID-19 elements, demographic and environment features, spatial and temporal variables, and the obtained predictions were assessed under different situations for future COVID-19 diffusion. Wei and Hansen [34] constructed a combined demand structure for air passenger traffic in a hub-and-spoke system. The model includes different parameters such as service frequency, aircraft size, ticket price, flight distance, and number of spokes in the network. Alekseev and Seixas [35] proposed a multivariate neural forecasting model for air transport passenger in Brazilian. The proposed neural processing performs better the conventional econometric method. Hsiao and Hansen [36] provided an air passenger model that use demand assignment in one structure which uses both time series and cross-sectional panel data of air travel demand using three-level nested logit model.

Rajendran et al. [37] examined prediction of the demand for air taxi urban movement services in New York including several ride-connected factors and weather-connected factors employing popular machine algorithm tools such as logistic regression, ANN, RF, and gradient boosting. Gradient boosting provides better results. Qin et al. [38] suggested two new hybrid methods integrating seasonal-trend decomposition processes based on loess with echo state network enhanced by grasshopper optimization algorithm and adaptive boosting (Adaboost) approach separately to estimate monthly passenger movement in China. The experimental results show that the offered methods ensure higher accuracy compared to other forecasting methods. Vadlamani et al. [39] considered machine learning algorithms (five famous machine learning models: logistic regression, decision tree, support vector machine, Scope rules and random forest) to evaluate and forecast entry shapes of Southwest Airlines into several city pairs. It was revealed that both logistic regression and decision tree models were successful in forecasting the entry shapes.

In this study, three alternative methods are utilized for rigorous forecasting of air passenger flows in Turkey: ANN, SVR, and RF techniques. The main force for deciding these prediction approaches is that they are favourable, qualified, and effective machine learning tools due to their easy adaptability to dynamic data constructions.

3.3 METHODOLOGY

In this section, the employed machine learning models are defined as ANN, SVM, RF respectively. In addition, MLR and times series were included to compare and exhibit predictive success of the ANN, SVR, and RF approaches.

3.3.1 ANN

The artificial neural network (ANN) approach is principally based on interrelated biological neural networks, such as in the cognitive system of the human brain. ANN has an algorithm configuration which performs by inclusion interrelated and multi-layered nodes. Neural networks comprise a vast number of computational components (neurons) relating through weighted links. Scholars demonstrated a great attention in ANN since its learning complex knowledge patterns ability under restricted data structure. ANNs can be categorized into several types, such as supervised and unsupervised learning methods, and feedforward and feedback recall designs. In forecasting models, a back propagation neural network architecture is commonly utilized in neural network tools [41]. The fundamental indication of the backpropagation algorithm is to reduce total square error rates by backward propagating along the neural network. To augment the learning progression of the backpropagation algorithm, two factors of the algorithm that contain the learning degree and the momentum should be adapted depends on the degree of convergence [42] (Figure 3.1).

The success of ANN depends on competently adjusted network structure based on diverse parameter integration such as number of neurons, layers, iteration, learning

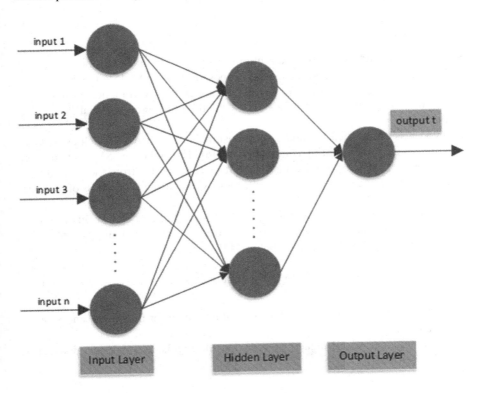

FIGURE 3.1 A general ANN architecture.

algorithm, and transfer function. One of the factors between these combinations is the determination of an appropriate transfer function. The utilized hyperbolic tangent sigmoid transfer function is specified in Equation 3.1:

$$f(s) = \frac{1 - e^{-s}}{1 + e^{-s}} \tag{3.1}$$

where s is the weighted input sum of the hidden layer, and $f(s)$ is the output of the hidden layer.

The general formula employed in the ANN building is given below [43].

$$y = f\left(\sum w_i x_i + b\right) \tag{3.2}$$

where
f = trasfer function, w = weights, b = bias.

For more scientific background of the technique, readers can obtain further knowledge at [44].

3.3.2 SVR

SVM was revealed as a strong data mining technique in classification, regression and pattern identification problems and proposed by Vapnik [45]. The SVM technique used for binary classification models although the SVM model involves general estimation problems, comprises a SVM method for regression that is the SVR method.

SVM was established to handle some insufficiencies of ANN, such as network building issues and overfitting complications [46]. The aim of the SVR is to explore a function that diverges at most insensitive loss function (ε) from the real outputs. The overfitting issue is reduced by simultaneously increasing the flatness of the function and reducing the error.

Support vector regression (SVR) is applied when linear models cannot be built with the existing data. The general form utilized for SVR in nonlinear regression models is specified mathematically below [47].

$$f(x) = w \times \varphi(x) + b \tag{3.3}$$

where $\varphi(x)$ symbolizes the kernel transformation model for the inputs, and w and b are parameters

The coefficients are designed by reducing the regularized risk function that is specified below,

$$R(f) = C\frac{1}{n}\sum_{i=1}^{n} L_\varepsilon\big(y_i, f(x_i)\big) + \frac{1}{2}\|w\|^2 \tag{3.4}$$

where the parameter ε is the amount of tolerance.

$$L_{\varepsilon}\left(y,f\left(x\right)\right)=\begin{cases} 0, & \left|y-f\left(x\right)\right|<\varepsilon \\ \left|y-f\left(x\right)\right|-\varepsilon, & \left|y-f\left(x\right)\right|\geq\varepsilon \end{cases} \tag{3.5}$$

In the described equation, $L_{\varepsilon}(y, f(x))$ is characterized a ε insensitive loss function. And after all the required mathematical operations can be accessed from [48], the model was transformed to dual Lagrangian form and finally utilizing Karush-Kuhn-Tucker statement, the following regression model is constructed by:

$$f\left(x\right)=\left(\beta_i-\beta_i^*\right)K\left(x_i,x_j\right)+b \tag{3.6}$$

$K(x_i, x_j)$ is a kernel function whose rate matches the inner product of two vectors, x_i and x_j, in the feature space $\varphi(x_i)$ and $\varphi(x_j)$. The most prevalent kernel functions given as radial basis kernel, sigmoidal kernel, polynomial kernel, and linear kernel.

The polynomial kernel with an order of n and constant of c_1 and c_2 can be expressed in Equation 3.7:

$$K\left(x_i,x_j\right)=\left(c_1 x_i x_j + c_2\right)^n \tag{3.7}$$

The success of the SVR based on a good establishment of the hyper-limitations (C, ε) and the kernel factors (n). Correct definition of all three parameters considerably affects the prediction correctness of SVR instruments [46].

3.3.3 RF

Random Forest was proposed as a new method in 2001 by Leo Breiman [49]. It is a widely utilized machine learning algorithm applied for both classification and regression problems, incorporating the output of several decision trees arrive at a final result. It is the developed model of the Bagging method by adding the randomness feature. In RF algorithms, more trees in the forest provide higher accuracy and prevent overfitting. Due to its ability to handle dataset stack and ensure model accuracy, RF is widely implemented in various fields.

The concept of random forest is to develop the variance decline of the Bagging method by decreasing the correlation among trees without growing the variance too much. It is accomplished by random selection of input factors in tree expansion. In the working principle of RF algorithms the number of trees and the number of factors are two important parameters [50] (Figure 3.2).

The mathematically description of RF is specified below [51].

$$\hat{f}\left(x\right)=\frac{1}{J}\sum_{j=1}^{J}\hat{h}_j\left(x\right) \tag{3.8}$$

where
  x is the input vector,
  $\hat{h}_j(x)$ = the predicted output of the jth regression tree $\left(j=1,...,J\right)$,
  $\hat{f}_j(x)$ = the unweighted average value of the predictions of each regression trees.

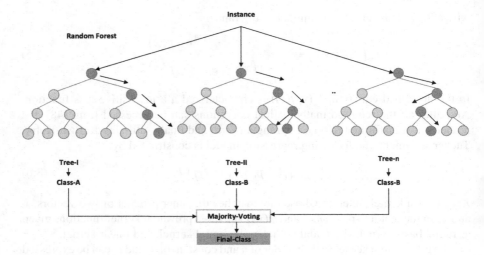

FIGURE 3.2   RF algorithm.

The sum of the squared residuals calculated when creating regression trees can be connected with each input attribute in each split in a regression tree and utilized to provide the variance extent. It is ensured that the features with the highest mean decrease in variance are the most significant.

3.4   APPLICATION OF THE METHODS: THE CASE OF TURKEY'S AIR PASSENGER DEMAND FORECASTING

In the presented approach, we firstly examine Turkey's air passenger demand patterns historically, and then exploit numerous machine learning tools for an efficient forecasting objective. After implementing machine learning instruments, we evaluate their performance by comparing the predicted values with actual scores.

In this study, the air passenger demand forecasting has been implemented in Turkey for the period 2000–2022. Based on a comprehensive review of the literature, reliable data sources and discussions with professionals and academics with relevant knowledge and proficiency in airway projections, we initially specified a set of predictor factors which cover as population, exchange rates, GDP per capita, import/export rates, tourism revenue per year, number of flights etc. which are among the most significant factors of air passenger for demand projections.

The dataset comprises 23 yearly observations. As seen in Figure 3.3, the number of passengers has a gradual increasing trend except for one minor drop and one major drop. The first drop is connected to the political, economic and social tensions affected to the tourism activities and then the second dramatic decline related to outbreak of COVID-19 which was publicized as a pandemic by World Health Organization in March 2020.

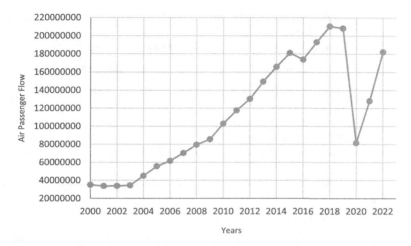

FIGURE 3.3 Air passenger demand from 2000 to 2022.

To implement an accurate prediction model, we need a normalization to initially convert the entire dataset into a rescaling [0,1] range to avoid excessive deviations in different data structures using the following equation:

$$x_{norm} = \frac{x - x_{min}}{x_{max} - x_{min}} \tag{3.9}$$

In order to compute the precision of the fitted models in the research, we evaluated different performance criteria such as root mean square error (RMSE), mean absolute percentage error (MAPE), and mean square error (MSE) by the following formulas in Table 3.4:

where N is the observation number, y_i is the actual observations, and $\hat{y}$ is the predicted data. The most widely utilized scale of accuracy for testing models is listed in Table 3.5 [52].

Correlation analysis is a guide to establish whether or not there is a relationship between factors/datasets, and what the degree of such a relationship might be. The relationship of all factors with the number of passengers was tested by correlation analysis. The summary statistics of related variables are shown in Tables 3.6 and 3.7, and the scatter plots are given in Figure 3.4.

TABLE 3.4
Different Performance Criteria

$$RMSE = \sqrt{\frac{\sum_{i=1}^{N}(y_i - \hat{y}_i)^2}{N}} \qquad MAPE = \frac{1}{N}\sum_{i=1}^{N}\left|\frac{y_i - \hat{y}_i}{y_i}\right| \times 100\% \qquad MSE = \frac{\sum_{i=1}^{N}(y_i - \hat{y}_i)^2}{N} \qquad MAE = \frac{1}{N}\sum_{i=1}^{N}|y_i - \hat{y}_i|$$

TABLE 3.5
Model Evaluation in Terms of MAPE (%) Levels

Prediction Power	MAPE (%)
Highly accurate prediction	<10
Good prediction	10–20
Reasonable prediction	20–50
Weak and inaccurate prediction	>50

TABLE 3.6
Basic Statistics of Variables

Factors	Unit	Mean	Min Value	Max Value	SD
Population	Person	74,978,215	65,603,160	85,279,553	6,405,273
Gross Domestic Percentage per Capita	USD ($)	10.282	7.961	12.489	1.246
Tourism Revenue	$1,000	24,109,678	11,254,892	46,284,907	8,688,901
Exports	$1,000	127,610,882	27,774,906	254,171,899	60,932,654
Imports	$1,000	180,889,322	12,105,678	363,710,986	86,727,841
Exchange rate	(Turkish Lira/USD)	3.95	1.299	16.604	3.95
Flight Traffic	Million	959,224	374,987	1,556,417	434,086
Air Passenger	Person	111,258,206	33,620,448	210,498,164	61,380,309

TABLE 3.7
Correlation Analysis

	Pop.	GDP per Capita	Tourism R.	Exports	Imports	Exchange Rate	Flight Traffic
Pearson correlation coefficient	0.694*	0.323	0.650*	0.602*	0.516*	0.360	0.988*
p-value	0.02	0.206	0.005	0.011	0.034	0.156	0.000

* Correlation is significant at the 0.05 level.

Initially, we performed a correlation analysis using Minitab statistical software on the related data set, taking the number of air passengers as the dependent variable, and the remaining factors as independent variables. Five of seven factors were associated with the number of passengers. These factors were obtained as population, tourism revenues, imports, exports, and flight traffic.

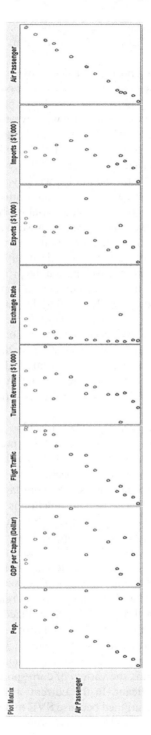

FIGURE 3.4   Scatter plot of air passenger vs. other indicators.

In Figure 3.4 scatter plot of yearly average population, GDP per capita, Tourism revenue, exchange rate (Turkish Lira/USD), exports, imports and number of air passenger were presented.

Figure 3.4 shows that there is a positive relation between population and number of air passenger values. The Pearson correlation coefficient of 0.694 indicates that number of air passenger is affected by population and should be employed as a variable in the prediction model. The effect of flight traffic was investigated with the scatter plot between flight traffic and the number of air passengers, and supported a linear relationship, as seen in Table 3.6. The Pearson correlation coefficient between the values of flight traffic and number of airline passengers is 0.988, which supports a significant positive relationship between them. Similarly, when we checked the association between tourism revenue, exports, imports and the output variable, air passenger, there is a positive association between related variables but remaining two factors which are GDP per capita and exchange rate show a weak association on the number of air passengers and their Pearson correlation coefficient value is approx. 0.323.

To conduct accurate analysis and forecast with multiple variables, we have normalized the related indicators and then we applied the proposed ANN configuration. Throughout ideal ANN architecture, various parameter combinations have been conducted and then reached the most suitable building design as follows.

The sample data, the flight passenger demand of Turkey between 2000 and 2022, were collected from the Civil Aviation Administration of Turkey and Turkish Statistical Institute. The 21 observations from 2000 to 2020 are utilized for model construction phase, and the remaining statistics from 2021 to 2022 are considered to assess the out-sample prediction performance of the models. Accordingly, while applying machine learning algorithms, the data set is separated into two parts; the training set and the test set.

In this study, we designated three prominent machine learning algorithms which are ANN, SVR and RF techniques to recognize the best prediction technique to project the yearly demand of air passenger in Turkey. Weka is open-source software, applied to evaluate performance of each approach.

The construction the ANN model covers five input layer, three hidden layer and one output layer. In the hidden layer, sigmoid function is employed as an activation function while linear function is utilized in the output layer. Levenberg-Marquardt optimization algorithm is utilized to evaluate the weights of neurons.

After altering the number of nodes (1–7) throughout the testing of the model, it was decided to employ three nodes at the hidden layer and learning rate is 0.2 and momentum value is 0.3 where the least mean square error (MSE) scale arisen at this value for the training data set. The MSE values were at minimum for both the training and testing data sets when three nodes were utilized in the hidden layer. The obtained relative absolute error rate is 5.303% and root relative squared error rate is 6.972% and the correlation coefficient is 0.9978 (Figure 3.5).

The main aim is to decrease the ε insensitive errors on training set when proposing the SVM configuration. Therefore, the variety of corresponding kernel functions and insensitive coefficients is significant. In the current study, the SVR method with polynomial kernel equation was utilized because SVR model with polynomial kernel model beats other SVR simulations. To increase progression of the forecasting success of SVR with polynomial kernel equation, we created the values of regularization

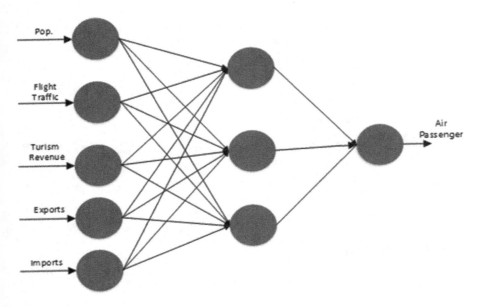

FIGURE 3.5 The developed ANN model.

parameter ($C = 4$) and insensitive coefficient ($\varepsilon = 0.001$) to lessen the error in the test dataset. The obtained relative absolute error rate is 6.373% and root relative squared error rate is 10.034% and the correlation coefficient is 0.996 (Figure 3.6).

One of the advanced machine learning approaches used for demand forecasting is the RF algorithm. The established tree configuration in the RF model is demonstrated in Figures 3.7 and 3.8. The number of iterations is set to 500 and any randomly

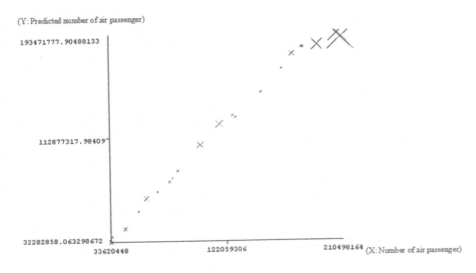

FIGURE 3.6 SVR model.

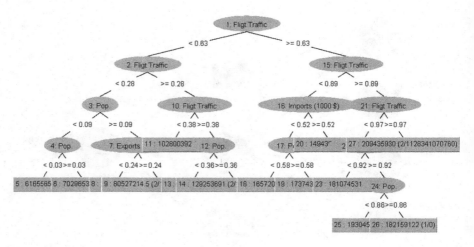

FIGURE 3.7 RF model.

```
RandomTree
==========

Fligt Traffic < 0.63
|    Turism Revenue ($1,000) < 0.28
|    |    Fligt Traffic < 0.12 : 61655659 (2/0)
|    |    Fligt Traffic >= 0.12 : 81616140 (2/0)
|    Turism Revenue ($1,000) >= 0.28
|    |    Pop. < 0.3
|    |    |    Turism Revenue ($1,000) < 0.38 : 102800392 (2/0)
|    |    |    Turism Revenue ($1,000) >= 0.38 : 79438289 (1/0)
|    |    Pop. >= 0.3
|    |    |    Fligt Traffic < 0.53 : 117620469 (1/0)
|    |    |    Fligt Traffic >= 0.53 : 128155762 (1/0)
Fligt Traffic >= 0.63
|    Fligt Traffic < 0.83
|    |    Fligt Traffic < 0.71 : 149430421 (1/0)
|    |    Fligt Traffic >= 0.71 : 165720234 (3/0)
|    Fligt Traffic >= 0.83
|    |    Fligt Traffic < 0.96
|    |    |    Fligt Traffic < 0.92 : 181074531 (1/0)
|    |    |    Fligt Traffic >= 0.92 : 193045343 (2/0)
|    |    Fligt Traffic >= 0.96 : 210498164 (1/0)

Size of the tree : 21
```

FIGURE 3.8 A randomly chosen decision tree configuration in RF algorithm.

chosen decision tree structure is given to show the progression of the proposed RF configuration. The obtained relative absolute error rate is 6.069% and root relative squared error rate is 8.906% and the correlation coefficient is 0.9962.

3.5 COMPARISON OF THE OBTAINED RESULTS

The comparison of results for future projection studies acquired by using ARIMA, MLR and machine learning algorithms is given below. Diverse performance criteria such as MAPE, RMSE, and MAE are employed to assess the forecasting performance of all three advanced models. Table 3.8 provides the analytical results of the several forecasting methods. Machine learning algorithms outperforms the traditional forecasting methods. The experimental results show that the ANN approach has the lowest MAPE (2.20%), RMSE (4,245,041), and MAE (2,783,395). SVR gives better results than RF methodology by a small margin with 3.00% MAPE, RMSE (6,295,248), and MAE (3,664,316) values. The results of MAPE, RMSE, and MAE indicate that the employed data mining tools, can effectively improve prediction accuracy.

In order to assess the performance of the utilized machine learning tools, also we perform the traditional and popular approaches including MLR and ARIMA (0,1,0) on the related dataset. The analysis results indicate that the ARIMA (0,1,0) has the highest MAPE (15.8%), RMSE (30,781,222) and MAE (13,494,924). MLR method gives better results than ARIMA with 4.1% MAPE, 5,086,323 RMSE and 3,933,056 MAE values. However, performance indicators of both ARIMA and MLR models are also worse than the recently developed machine learning tools. It can clearly be seen that the MAPE, RMSE, and MAE results point out that ANN, SVR and RF can provide more reliable and effective prediction accuracy.

Table 3.9 summarizes the simulation and forecasting results obtained by ANN, SVR, and RF models. Actual and prediction values for the proposed machine learning approaches are demonstrated in Figure 3.9.

The MAPE values of ANN, SVR, and RF models implemented in the model building part (2000–2020) are 2.20%, 3.00%, and 3.21%, respectively. Regarding the results of the model testing part of the forecast (2021–2022), the MAPE values are 2.35%, 0.62%, and 5.90%, respectively. As can be seen in Table 3.9, the employed

TABLE 3.8
The Performance Comparison with Different Forecasting Methods

Forecasting Method/Data Set	Training Data			Test Data		
	MAPE	RMSE	MAE	MAPE	RMSE	MAE
ANN	0,022	4,245,041	2,783,395	0,023	3,497,707	3,461,915
SVR	0,030	6,295,248	3,664,316	0,006	1,097,308	806,597
RF	0,032	4,526,959	2,840,888	0,059	10,655,002	7,580,053
MLR	0,041	5,086,323	3,933,056	0,018	2,685,655	2,599,896
ARIMA (0,1,0)	0,158	30,781,222	13,494,924	0,2899	44,440,439	44,280,510

TABLE 3.9

Comparison of Models for Air Passenger Demand

Periods	Actual Value	ANN Predicted	Error (%)	SVR Predicted	Error	RF Predicted	Error
2000	34,972,534	35,695,981	2,06	36,257,399	3.67	36,730,411	5.03
2001	33,620,448	33,160,105	−1,36	33,683,078	0.18	34,633,424	3.01
2002	33,783,892	33,749,885	−0,10	33,443,511	−1,01	34,633,424	2.51
2003	34,443,655	34,680,074	0,68	32,282,858	−6,27	34,747,793	0.88
2004	45,057,371	43,402,434	−3,67	42,221,936	−6,29	42,449,274	−5,78
2005	55,572,426	52,759,974	−5,06	55,883,414	0.55	52,869,039	−4,86
2006	61,655,659	62,084,704	0,69	65,459,011	6.16	63,126,914	2.38
2007	70,296,532	69,923,225	−0,53	70,636,912	0.48	71,811,456	2.15
2008	79,438,289	80,300,672	1,08	78,275,256	−1,46	79,918,091	0,60
2009	85,508,508	85,663,635	0,18	86,829,568	1.54	84,476,501	−1,20
2010	102,800,392	102,132,347	−0,64	107,027,620	4.11	98,464,204	−4,21
2011	117,620,469	119,711,362	1,77	123,165,212	4.71	122,416,099	4.07
2012	130,351,620	125,610,312	−3,63	128,232,227	−1,62	129,299,254	−0,80
2013	149,430,421	145,217,050	−2,81	147,767,644	−1,11	149,164,008	−0,17
2014	165,720,234	164,590,498	−0,68	165,616,423	−0,06	167,345,071	0.98
2015	181,074,531	185,426,381	2,40	182,528,843	0.80	179,654,027	−0,78
2016	173,743,537	179,067,630	3,06	177,094,022	1.2	173,743,537	0
2017	193,045,343	186,345,117	−3,47	184,041,864	−4,66	188,001,063	−2,61
2018	210,498,164	196,587,498	−6,60	189,820,966	−9,82	204,948,996	−2,63
2019	208,373,696	206,189,969	−1,04	193,471,777	−7,15	203,081,902	−2,53
2020	81,616,140	76,220,232	−6,61	81,333,780	−0,34	98,159,622	20.26
MAPE (%) (2000–2020)			2,20		3.00		3,21
Out of Sample							
2021	128,155,762	132,116,778	0,03	129,706,326	0,012	143,223,931	0,12
2022	182,159,122	185,121,935	0,016	182,221,752	0,001	182,067,185	0,001
MAPE (%) (2021–2022)			2,35		0.62		5,90

machine learning algorithms have a higher forecast accuracy in both model building and model testing phases.

Forecasting values for suggested models are also exemplified in Figure 3.9.

Figure 3.10 demonstrates the error rate of the forecasting models over the years.

Due to the successful results offered by ANN, the projection for the next 10 years was made using ANN configuration and given in Figure 3.11. The Figure 3.11 displays the future projections for the next years (2023–2032). The obtained results show that the air passenger demand of Turkey will reach approximately 300 million

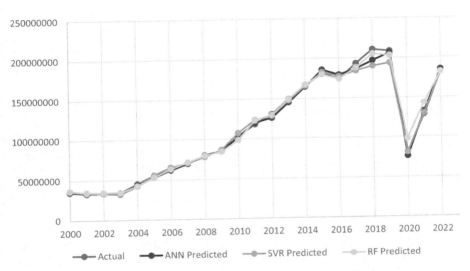

FIGURE 3.9 Air passenger demand forecasting with the proposed machine learning algorithms.

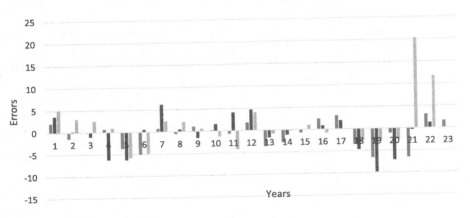

FIGURE 3.10 Error rates of the forecasting models.

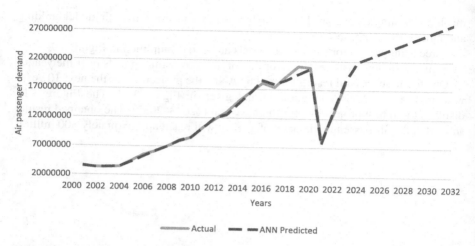

FIGURE 3.11   The predicted air passenger demand from 2023–2032.

(290,897,125) by 2032. More precisely, it implies that Turkey's airline passenger flows will nearly double in less than 20 years (between 2013 and 2032, from 149 million to 300 million).

3.6 CONCLUSION

Due to the different pattern structure in data and volatile dynamics of air transportation system, accurately predicting of passenger flows can be a daunting task. The use of advanced forecasting techniques provides numerous advantages in order to see the future today, to take the necessary measures in case of a setback, to minimize the loss in terms of profitability and to maintain its presence in the market despite its competitors.

Comprehensively designed analytical analysis results guide policy makers, managers and experts working in this field in determining new flight routes and capacity building activities by airlines as well as longer-term management side decisions such as risk sharing and public economics.

In this study, we evaluated the recently popular machine learning algorithms ANN, SVR, and RF under the umbrella of providing successful and long-term projection using the data of air passenger flows in Turkey in the last 23 years. We assessed the success of the methods by controlling the error rates. The lowest error margin provides the highest accuracy and the highest success rate. The generated model and the estimated values can be useful to many actors in the aviation industry.

According to the correlation analysis, the most relevant factors were determined as population, tourism income, flight traffic, exports and imports, and then introduced as input variables in ANN, SVR, and RF algorithm methods. ANN outperforms other machine learning tools in terms of MAPE scale with 2.21%, SVR 3.00%, and RF provides 3.21% subsequently. These powerful artificial intelligent tools are compared to traditional forecasting methods such as MLR and ARIMA to show their performance. All three advanced forecasting methods were well-formed and had less than 5% error in estimating actual passenger demand values; of the conventional methods, MLR provided a better model structure than ARIMA model and both offered lower prediction accuracy/success than machine learning tools. In fact, although regression analysis is also a machine learning tool, it has been accepted as one of the classical methods in this study with the idea of fundamental working principles.

In future studies, new variables can be added to the prediction model in combination with other advanced methods with their integrated performed to attain greater accuracy. In this way, traditional methods and advanced methods can be evaluated together and presented with all their weaknesses and strengths. Various situations such as low, medium and high air passenger demand scenarios should be developed in order to overcome unexpected situations such as any economic, political, social or energy-related crisis or unusual globally significant situations such as pandemics.

REFERENCES

[1] Castiglioni, M., Gallego, Á. and Galán, J. L. "The virtualization of the airline industry: A strategic process," *J. Air Transp. Manag.*, vol. 67, pp. 134–145, Mar. 2018, doi: 10.1016/J.JAIRTRAMAN.2017.12.001

[2] Baker, D. M. A. "Service quality and customer satisfaction in the airline industry: A comparison between legacy airlines and low-cost airlines," *Am. J. Tour. Res.*, vol. 2, no. 1, pp. 67–77, 2013, doi: 10.11634/216837861403317

[3] Hofmann, E. and Rutschmann, E. "Big data analytics and demand forecasting in supply chains: A conceptual analysis," *Int. J. Logist. Manag.*, vol. 29, no. 2, pp. 739–766, 2018, doi: 10.1108/IJLM-04-2017-0088/FULL/XML

[4] Gössling, S. and Humpe, A. "The global scale, distribution and growth of aviation: Implications for climate change," *Glob. Environ. Chang.*, vol. 65, p. 102194, Nov. 2020, doi: 10.1016/J.GLOENVCHA.2020.102194

[5] Mazareanu, E. "Revenue of airlines worldwide 2003–2022 I Statista," 2021. https://www.statista.com/statistics/278372/revenue-of-commercial-airlines-worldwide/ (Accessed Mar. 15, 2023).

[6] IATA, "IATA: Below-trend but still solid air passenger growth in 2019," 2019. https://www.iata.org/en/publications/economics/?EconomicsL2=146&Ordering=DateDesc&Search=&page=3 (Accessed Mar. 15, 2023).

[7] Zhang, A. "Analysis of an international air-cargo hub: The case of Hong Kong," *J. Air Transp. Manag.*, vol. 9, no. 2, pp. 123–138, Mar. 2003, doi: 10.1016/S0969-6997(02)00066-2

[8] Liehr, M, Größler, A, Klein, M. and Milling, P. M. "Cycles in the sky: Understanding and managing business cycles in the airline market," *Syst. Dyn. Rev.*, vol. 17, no. 4, pp. 311–332, Dec. 2001, doi: 10.1002/SDR.226

[9] Helmreich, R. L., Merritt, A. C. and Wilhelm, J. A. "The evolution of crew resource management training in commercial aviation," *Hum. Error Aviat.*, pp. 275–288, Jul. 2017, doi: 10.4324/9781315092898-15

[10] Suryani, E., Chou, S. Y. and C. H. Chen, "Air passenger demand forecasting and passenger terminal capacity expansion: A system dynamics framework," *Expert Syst. Appl.*, vol. 37, no. 3, pp. 2324–2339, Mar. 2010, doi: 10.1016/J.ESWA.2009.07.041

[11] Profillidis, V. A. "Econometric and fuzzy models for the forecast of demand in the airport of Rhodes," *J. Air Transp. Manag.*, vol. 6, no. 2, pp. 95–100, Apr. 2000, doi: 10.1016/S0969-6997(99)00026-5

[12] Kitsou, S. P., Koutsoukis, N. S., Chountalas, P. and Rachaniotis, N. P. "International passenger traffic at the Hellenic airports: Impact of the COVID-19 pandemic and mid-term forecasting," *Aerospace* vol. 9, no. 3, p. 143, Mar. 2022, doi: 10.3390/AEROSPACE9030143

[13] Boonekamp, T., Zuidberg, J. and Burghouwt, G. "Determinants of air travel demand: The role of low-cost carriers, ethnic links and aviation-dependent employment," *Transp. Res. Part A Policy Pract.*, vol. 112, pp. 18–28, Jun. 2018, doi: 10.1016/J.TRA.2018.01.004

[14] Solvoll, G., Mathisen, T. A. and Welde, M. "Forecasting air traffic demand for major infrastructure changes," *Res. Transp. Econ.*, vol. 82, p. 100873, Oct. 2020, doi: 10.1016/J.RETREC.2020.100873

[15] Jierula, A., Wang, S., Oh, T. M. and P. Wang. "Study on accuracy metrics for evaluating the predictions of damage locations in deep piles using artificial neural networks with acoustic emission data," *Appl. Sci.* vol. 11, no. 5, p. 2314, Mar. 2021, doi: 10.3390/APP11052314

[16] General Directorate Of State Airports, "Havayolu Sektör Raporları," 2020. https://www.dhmi.gov.tr/Sayfalar/HavaYoluSektorRaporlari.aspx (Accessed Mar. 19, 2023).

[17] Qin, L., Li, W. and Li, S., "Effective passenger flow forecasting using STL and ESN based on two improvement strategies," *Neurocomputing*, vol. 356, pp. 244–256, Sep. 2019, doi: 10.1016/J.NEUCOM.2019.04.061

[18] Jin, F., Li, Y., Sun, S. and Li, H. "Forecasting air passenger demand with a new hybrid ensemble approach," *J. Air Transp. Manag.*, vol. 83, p. 101744, Mar. 2020, doi: 10.1016/J.JAIRTRAMAN.2019.101744

[19] Kim, S. and Shin, D. H. "Forecasting short-term air passenger demand using big data from search engine queries," *Autom. Constr.*, vol. 70, pp. 98–108, Oct. 2016, doi: 10.1016/J.AUTCON.2016.06.009

[20] Gelhausen M. C., Berster, P. and Wilken, D. "A new direct demand model of long-term forecasting air passengers and air transport movements at German airports," *J. Air Transp. Manag.*, vol. 71, pp. 140–152, Aug. 2018, doi: 10.1016/J.JAIRTRAMAN.2018.04.001

[21] Gunter, U. and Zekan, B. "Forecasting air passenger numbers with a GVAR model," *Ann. Tour. Res.*, vol. 89, p. 103252, Jul. 2021, doi: 10.1016/J.ANNALS.2021.103252

[22] Tang, H. et al., "Airport terminal passenger forecast under the impact of COVID-19 outbreaks: A case study from China," *J. Build. Eng.*, vol. 65, p. 105740, Apr. 2023, doi: 10.1016/J.JOBE.2022.105740

[23] Li Long, C., Guleria, Y. and Alam, S. "Air passenger forecasting using Neural Granger causal Google trend queries," *J. Air Transp. Manag.*, vol. 95, p. 102083, Aug. 2021, doi: 10.1016/J.JAIRTRAMAN.2021.102083

[24] Sun, S., Lu, H., Tsui, K. L. and Wang, S. "Nonlinear vector auto-regression neural network for forecasting air passenger flow," *J. Air Transp. Manag.*, vol. 78, pp. 54–62, Jul. 2019, doi: 10.1016/J.JAIRTRAMAN.2019.04.005

[25] Xie, G., Wang, S., and Lai, K. K. "Short-term forecasting of air passenger by using hybrid seasonal decomposition and least squares support vector regression approaches," *J. Air Transp. Manag.*, vol. 37, pp. 20–26, May 2014, doi: 10.1016/J.JAIRTRAMAN.2014.01.009

[26] Kağan Albayrak, M. B., Özcan, I. Ç., Can, R. and Dobruszkes, F., "The determinants of air passenger traffic at Turkish airports," *J. Air Transp. Manag.*, vol. 86, p. 101818, Jul. 2020, doi: 10.1016/J.JAIRTRAMAN.2020.101818

[27] Xiao, Y., Liu, J. J., Hu, Y., Wang, Y., Lai, K. K. and S. Wang, "A neuro-fuzzy combination model based on singular spectrum analysis for air transport demand forecasting," *J. Air Transp. Manag.*, vol. 39, pp. 1–11, Jul. 2014, doi: 10.1016/J.JAIRTRAMAN.2014.03.004

[28] Suh, D. Y. and Ryerson, M. S. "Forecast to grow: Aviation demand forecasting in an era of demand uncertainty and optimism bias," *Transp. Res. Part E Logist. Transp. Rev.*, vol. 128, pp. 400–416, Aug. 2019, doi: 10.1016/J.TRE.2019.06.016

[29] Carmona-Benítez, R. B., Nieto, M. R. and Miranda, D. "An Econometric Dynamic Model to estimate passenger demand for air transport industry," *Transp. Res. Procedia*, vol. 25, pp. 17–29, Jan. 2017, doi: 10.1016/J.TRPRO.2017.05.191

[30] Fildes, R., Wei, Y. and Ismail, S. "Evaluating the forecasting performance of econometric models of air passenger traffic flows using multiple error measures," *Int. J. Forecast.*, vol. 27, no. 3, pp. 902–922, Jul. 2011, doi: 10.1016/J.IJFORECAST.2009.06.002

[31] Grubb, H. and Mason, A. "Long lead-time forecasting of UK air passengers by Holt–Winters methods with damped trend," *Int. J. Forecast.*, vol. 17, no. 1, pp. 71–82, Jan. 2001, doi: 10.1016/S0169-2070(00)00053-4

[32] Nieto, M. R. and Carmona-Benítez, R. B. "ARIMA + GARCH + Bootstrap forecasting method applied to the airline industry," *J. Air Transp. Manag.*, vol. 71, pp. 1–8, Aug. 2018, doi: 10.1016/j.jairtraman.2018.05.007

[33] Dey Tirtha, S., Bhowmik T., and Eluru, N. "An airport level framework for examining the impact of COVID-19 on airline demand," *Transp. Res. Part A Policy Pract.*, vol. 159, pp. 169–181, May 2022, doi: 10.1016/J.TRA.2022.03.014

[34] Wei, W. and Hansen, M. "An aggregate demand model for air passenger traffic in the hub-and-spoke network," *Transp. Res. Part A Policy Pract.*, vol. 40, no. 10, pp. 841–851, Dec. 2006, doi: 10.1016/J.TRA.2005.12.012

[35] Alekseev, K. P. G. and Seixas, J. M. "A multivariate neural forecasting modeling for air transport – Preprocessed by decomposition: A Brazilian application," *J. Air Transp. Manag.*, vol. 15, no. 5, pp. 212–216, Sep. 2009, doi: 10.1016/J.JAIRTRAMAN.2008.08.008

[36] Hsiao, C. Y. and Hansen, M. "A passenger demand model for air transportation in a hub-and-spoke network," *Transp. Res. Part E Logist. Transp. Rev.*, vol. 47, no. 6, pp. 1112–1125, Nov. 2011, doi: 10.1016/J.TRE.2011.05.012

[37] Rajendran, S., Srinivas S. and Grimshaw, T. "Predicting demand for air taxi urban aviation services using machine learning algorithms," *J. Air Transp. Manag.*, vol. 92, p. 102043, May 2021, doi: 10.1016/J.JAIRTRAMAN.2021.102043

[38] Qin, L., Li, W. and Li, S. "Effective passenger flow forecasting using STL and ESN based on two improvement strategies," *Neurocomputing*, vol. 356, pp. 244–256, Sep. 2019, doi: 10.1016/J.NEUCOM.2019.04.061

[39] Vadlamani, S. L., Shafiq, M. O. and Baysal, O. "Using machine learning to analyze and predict entry patterns of low-cost airlines: A study of Southwest Airlines," *Mach. Learn. with Appl.*, vol. 10, p. 100410, Dec. 2022, doi: 10.1016/J.MLWA.2022.100410

[40] Sun, S., Lu, H., Tsui, K. L. and Wang, S. "Nonlinear vector auto-regression neural network for forecasting air passenger flow," *J. Air Transp. Manag.*, vol. 78, pp. 54–62, Jul. 2019, doi: 10.1016/j.jairtraman.2019.04.005

[41] Pankratz, A. *Forecasting with univariate Box-Jenkins models: Concepts and cases*, vol. 224. Hoboken, NJ: John Wiley & Sons, 2009.

[42] Wang, A. J. and Ramsay, B. "A neural network based estimator for electricity spot-pricing with particular reference to weekend and public holidays," *Neurocomputing*, vol. 23, no. 1–3, pp. 47–57, 1998, doi: 10.1016/S0925-2312(98)00079-4

[43] Cayir Ervural, B. "A combined methodology for evaluation of electricity distribution companies in Turkey," *J. Intell. Fuzzy Syst.*, vol. 38, no. 1, pp. 1059–1069, Jan. 2020, doi: 10.3233/JIFS-179468

[44] Hagan, M. T. and Menhaj, M. B. "Training feedforward networks with the Marquardt algorithm," *IEEE Trans. Neural Networks*, vol. 5, no. 6, pp. 989–993, 1994, doi: 10.1109/72.329697

[45] Vapnik, V. N. *Statistical learning theory*. New York: Wiley, 1998.

[46] Kavousi-Fard, A., Samet, H. and Marzbani, F. "A new hybrid modified firefly algorithm and support vector regression model for accurate short term load forecasting," *Expert Syst. Appl.*, vol. 41, no. 13, pp. 6047–6056, Dec. 2014, doi: 10.1016/j.eswa.2014.03.053

[47] Li, J., Boonaert, J., Doniec, A. and Lozenguez, G. "Multi-models machine learning methods for traffic flow estimation from Floating Car Data," *Transp. Res. Part C Emerg. Technol.*, vol. 132, p. 103389, Nov. 2021, doi: 10.1016/J.TRC.2021.103389

[48] Beyca, O. F., Ervural, B. C., Tatoglu, E., Ozuyar, P. G. and Zaim, S. "Using machine learning tools for forecasting natural gas consumption in the province of Istanbul," *Energy Econ.*, vol. 80, pp. 937–949, May 2019, doi: 10.1016/J.ENECO.2019.03.006

[49] Breiman, L. "Random forests," *Mach. Learn.*, vol. 45, no. 1, pp. 5–32, Oct. 2001, doi: 10.1023/A:1010933404324/METRICS

[50] Archer, K. J. and Kimes, R. V. "Empirical characterization of random forest variable importance measures," *Comput. Stat. Data Anal.*, vol. 52, no. 4, pp. 2249–2260, Jan. 2008, doi: 10.1016/J.CSDA.2007.08.015

[51] Gradojevic, N., Kukolj, D., Adcock, R. and Djakovic, V. "Forecasting Bitcoin with technical analysis: A not-so-random forest?," *Int. J. Forecast.*, vol. 39, no. 1, pp. 1–17, Jan. 2023, doi: 10.1016/J.IJFORECAST.2021.08.001

[52] Lewis, C. D. "Industrial and business forecasting methods: A practical guide to exponential smoothing and curve fitting," *Butterworth Scientific*, 1982. https://iucat.iu.edu/iue/3760302 (Accessed Mar. 19, 2023).

4 Staff Scheduling Strategies in Aviation Industry

Kemal Alaykiran
Necmettin Erbakan University, Konya, Turkiye

4.1 INTRODUCTION

The aviation, or air transportation, industry is competitive, volatile, and cost-intensive. The volatility of the market is most commonly unpredictable and uncontrollable. As the world experienced the widespread, global COVID-19 pandemic, the aviation industry has been one of the most affected in terms of operations and profitability. Besides, economic downturns, changes in passenger behaviours, and global political issues also are of great significance for this industry. As a result of its highly competitive nature, controlling operational costs, choosing the correct market, and applying innovative marketing approaches are key to survival or increasing market share for companies in this industry. Since the price is not in the control of a single company or an alliance, optimizing operational costs should be one of the best ways to increase sales and the market share of a company. Fuel prices and crew costs are the most effective parameters in total operating cost. Since crude oil prices are not controllable by the industry, optimizing the second biggest cost parameter, crew cost, seems to be the most rational way for an airline to control its total cost.

The total staff of an airline comprises cockpit crew, cabin crew, ground staff, and management staff. The cockpit crew includes captains and officers who are the most important part of the whole crew due to their unique qualification of being able to fly aircraft. The cabin crew consists of flight attendants, stewards, and cabin mates, whose quantity and qualification are determined by security rules and individual company policies for better customer satisfaction. The ground staff are responsible for numerous ground operations which are crucial for the security and continuity of operations. Finally, the management staff are responsible for the operational, tactical, and strategic planning activities of an airline, like fleet planning, network planning, scheduling, marketing, advertising, amongst others. All these staff members must work in harmony for the company to meet its goals. Hence, one can easily rconclude that the aviation industry is a staff-intensive industry where expensive and special training is required. Also, one can easily mention that staffing in the aviation industry is cost-intensive since the crew cost is a major part of the total operations cost as mentioned. As a result of these two arguments, staff scheduling becomes of great importance for any company in this industry to optimize one of the most important cost issues that are controllable. The delicate optimization of the schedule of this staff requires developed mathematical models and analytical approaches rather than falling back upon human effort which does not guarantee optimal results.

In this study, a new methodology will be applied to the crew assignment problem using the Equitable Partitioning Method, to determine and assign the correct number of crew members to predetermined flights to generate groups of crew members as homogenous as possible to maintain customer satisfaction from the service proposed. Firstly, the crew members will be listed based on different attributes such as age, experience, gender, availability for different cabin classes (economy, business, first class), language competencies, etc. Secondly, a mixed-integer linear programming model will be generated to ensure that required number of crew members are assigned to each flight as a group with homogenous attribute values. The model generated will be solved using CPLEX for variations of the problem to simulate different real-life situations. The results will be evaluated from the solution time point of view.

The text will be constructed as follows. In the next section, the main framework of the staff scheduling environment of the aviation industry will be described and the literature studying this problem will be given in detail in the next section. In Section 4.3, the scheduling problem considered will be defined. In Section 4.4, the Equitable Partitioning Problem and the solution methodology will be explained. In Section 4.5, the mixed-integer linear programming model will be proposed. In Section 4.6, different cases will be solved using the model and the solutions will be analyzed. Finally, the text will be concluded with an analysis of the results and some future research remarks.

4.2   LITERATURE REVIEW

One of the easiest ways of reaching the most recent publications about a specific topic in the literature is to refer to a recent review article. Wen et al. (2021) carried out a comprehensive and detailed review of recent publications about airline crew scheduling. The literature review given in this study will be shaped using a similar methodology to the one used in Wen et al. (2021). For other review articles with different methodologies, one may also refer to Eltoukhy et al. (2017) and Deveci and Demirel (2018).

Wen et al. (2021) take the airline scheduling problem as a whole, addressing its large scale and high complexity, and partition this problem into Flight Scheduling, Fleet Assignment, Aircraft Maintenance Routing, and Crew Scheduling problem, besides, Crew Scheduling Problem is divided into two sub-categories as Crew Pairing, and Crew Assignment problems. Based on Wen et al. (2021), "…The crew pairing problem is to generate sufficient anonymous feasible pairings to satisfy all flight's manpower requirements while minimizing the total operating costs…" (Wen et al., 2021), where "…in the crew assignment/rostering problem, the pairings constructed in the crew pairing problem are connected to form monthly assignments/rosters to be assigned to specific crew members…" (Wen et al., 2021). In other words, the crew pairing problem is to assign the number of crew needed to a flight anonymously, whereas the crew assignment problem is to assign specific crew members to a flight at the required number. Based on these definitions, from the linear programming point of view, the crew assignment problem is harder to solve within a reasonable duration of time than the crew pairing problem.

Another diversification in the nature of publications is the crew type taken into consideration. Some publications consider only cockpit crew for scheduling, some others consider only cabin crew for scheduling, and only a couple of studies consider both crew types to be scheduled together. Due to some operational differences, these two crew types differ, and the complexity of the problem increases when both are considered together.

Yan and Tu (2002) developed a pure network model to solve the crew scheduling problem using the real data of a Taiwanese Airline and a series of computational experiments were carried out to show the effectiveness of the model presented (Yan & Tu, 2002).

In Yan et al. (2002), eight different scheduling models are introduced with the objective of minimizing total crewing costs for cabin crew in a Taiwanese Airline investigating the problem from different perspectives (Yan et al., 2002).

Wen et al. (2022) studied the crew pairing problem considering multi-class aircraft requirements. Stating the importance and the complexity of this problem, a mathematical model is generated to solve the problem for manpower utilization improvement and cost reduction objectives together (Wen et al., 2022).

The objective function of the crew scheduling problem is also another important parameter. The majority of the studies in the literature focus on cost minimization, where it is a fact that cost minimization and customer satisfaction are two contradictory objectives of an airline, especially from the cabin crew assignment point of view. Besides the fact that due to specific security regulations, there is a lower limit for the number of cabin crew to be assigned to each flight, the more the number of flight attendants assigned to a flight may guarantee, the better customer satisfaction, which may lead to higher customer loyalty.

Yildiz et al. (2017) considered the crew pairing problem with fatigue. Stating that fatigue is a major factor in accidents, they developed a mathematical model named "Three Process Model of Alertness" which claimed to be one of the most comprehensive fatigue models in literature. They used two real data sets to show the performance and the validity of the model proposed (Yildiz et al., 2017).

Badanik et al. (2021) also focused on the crew rostering problem, considering not only the total cost as an objective function, but also giving importance to crewmembers' fatigue as an issue. They concluded that rostering factors such as the timing of the flight, night shifts, short rest periods, consecutive early morning starts, extended duty periods, and consecutive duty days have a great impact on the crews' fatigue, which may result in unwanted outcomes such as misjudgements and slowed reaction times. Suggestions for how to avoid these undesired results are stated in the study. Badanik et al. (2021).

Another important issue in the crew scheduling problems in the aviation industry is the constraints of a linear mathematical model, most of which are the result of security concerns (i.e., compulsive resting periods). Even if most of the research in the literature considers satisfying these security issues, only a couple of them consider staff preferences, and individual capabilities (e.g., foreign languages spoken) for maximizing the job satisfaction of the staff, as well as customer expectations.

Quesnel et al. (2020) studied the crew pairing problem with language constraints and concluded that the proposed solution algorithm served as an effective solution method to the problem (Quesnel et al., 2020).

The planning horizon of the crew scheduling problem is also important. If the problem is a tactical level (semi-annually, quarterly, monthly, or weekly) decision-making problem, then the solution time for the mathematical model to be in terms of hours would be reasonable to achieve an optimal solution; however, if the planning period is operational level (daily, or even hourly), then the solution time for a concrete schedule should be in terms of seconds. Otherwise, all these efforts to develop a mathematical model would be meaningless, and the necessity of a faster solution method, like a heuristic approach, should be vital.

4.3   CREW ASSIGNMENT PROBLEM

As it is described in Wen et al. (2021), the airline scheduling problem comprises four components: Flight Scheduling, Fleet Assignment, Aircraft Maintenance Routing, and Crew Scheduling. Flight scheduling is the first step where the route and frequency decisions are made. Mostly these decisions arise from "Where should we fly?" and "How frequently should we fly there?" questions. From the management point of view, these decisions are strategic-level decisions where they have a very slight effect on customer satisfaction. Generally, a customer would not ask "Why does this airline not fly to this destination?", or "Why are these flights not so frequent?" Mostly these decisions are made based on the market type, demand fluctuations, business type, profitability, and cost issues, etc.

The next decision to be made should be the fleet assignment. As the routes and frequencies are determined, the next question arises which is "With which aircraft should we fly to this destination at this time?". Since the numbers and types of aircraft available for an airline is scarce, this problem addresses a strict constraint. Again, from the customers' point of view, this issue is not always a matter of satisfaction, as long as the aircraft is not too old or the cabin design is not repulsive for the customers. In other words, most of the customers would not ask "Why are we flying with this model of an aircraft family?".

The aircraft maintenance routing is completely a different issue which does not have any effect on customer satisfaction as long as required security measures are adhered to.

The last phase of airline scheduling is the crew scheduling problem, which, without any doubt, has the greatest effect on customer satisfaction with the service provided by the airline. Crew scheduling problems may be summarized as assigning the required number of crew members with the required amount of competencies to flights in order not only to provide minimum expectations of the International Civil Aviation Organization (ICAO) but also to maximize customer satisfaction while minimizing the total cost of operations. Since the total number of crew members is limited, the competencies of each crew member differ in different attributes solving this problem becomes very complex even ignoring special constraints such as language competencies, different cabin class types, etc.

In this study, the crew scheduling problem will be addressed by defining this problem as an equitable partitioning problem.

4.4   THE EQUITABLE PARTITIONING PROBLEM

The equitable partitioning problem may be defined as assigning a group of entities into equivalent groups considering different attributes to generate as homogenous groups as possible (O'Brien & Mingers, 1997). This problem definition is used frequently in the literature to create student teams (Holmberg, 2019; Krass & Ovchinnikov, 2010; Mingers & O'Brien, 1995; Rubin & Bai, 2015), or worker teams for specific projects (Erdeş, 2021; Fathian et al., 2017; Rahmanniyay et al., 2019; van de Water et al., 2007).

The methodology of solving the equitable partitioning problem will be demonstrated using an example. Consider there being 20 students with scores in maths, science, language, history, and arts, as seen in Figure 4.1. Then, consider that the objective of a teacher is to separate these students into four groups with five individuals in each, where the groups are expected to be as even as possible. Undoubtedly, the scores of these students are different for each of the attributes, and if one randomly assigns these students to the groups, some of the groups will have advantage over the others in terms of some attributes.

SCORES					
Student	Maths	Science	Language	History	Arts
1	95	97	66	51	56
2	94	51	75	49	71
3	95	77	55	86	84
4	66	83	66	66	71
5	56	52	47	76	61
6	70	79	96	84	87
7	90	84	97	63	55
8	58	79	47	95	97
9	83	80	95	64	65
10	45	67	73	71	51
11	80	87	54	89	86
12	80	75	54	70	75
13	41	96	77	68	43
14	94	99	71	42	72
15	92	51	58	55	88
16	88	46	93	43	78
17	87	60	66	44	82
18	84	60	66	97	98
19	90	58	45	40	97
20	90	59	100	95	85
Average	78.9	72	70.1	67.4	75.1

GROUP 1					
Students	Maths	Science	Language	History	Arts
7	90	84	97	63	55
8	58	79	47	95	97
9	83	80	95	64	65
12	80	75	54	70	75
15	92	51	58	55	88

GROUP 2					
Students	Maths	Science	Language	History	Arts
2	94	51	75	49	71
3	95	77	55	86	84
6	70	79	96	84	87
13	41	96	77	68	43
19	90	58	45	40	97

GROUP 3					
Students	Maths	Science	Language	History	Arts
4	66	83	66	66	71
5	56	52	47	76	61
14	94	99	71	42	72
17	87	60	66	44	82
20	90	59	100	95	85

GROUP 4					
Students	Maths	Science	Language	History	Arts
1	95	97	66	51	56
10	45	67	73	71	51
11	80	87	54	89	86
16	88	46	93	43	78
18	84	60	66	97	98

FIGURE 4.1   The data and the solution for the example.

In order to solve this problem, Mingers and O'Brien (1995) proposed the below linear mathematical model.

i	: students,
t	: teams,
j	: attributes,

Parameters:

a_{ij}	: attribute value for student i for attribute j (if it is binary, it is 0 or 1, otherwise the number).
m	: number of teams,
n	: number of individuals,
k	: number of attributes,

Decision Variables:

X_{it}	: 1 if individual i assign to team t, 0 otherwise,
d_{tj}^+	: the positive deviation value of attribute j of team t from goal,
d_{tj}^-	: the negative deviation value of attribute j of team t from goal,

$$\min Z = \sum_{t=1}^{m}\sum_{j=1}^{k}\left(d_{tj}^+ + d_{tj}^-\right) \qquad (4.1)$$

s.t.

$$\sum_{i=1}^{n}\left(X_{it}a_{ij}\right) - d_{tj}^+ + d_{tj}^- = \frac{\sum_{i=1}^{n}a_{ij}}{m} \qquad (\forall\, j, t) \qquad (4.2)$$

$$\sum_{t=1}^{m}X_{it} = 1 \qquad (\forall\, i) \qquad (4.3)$$

$$d_{tj}^+, d_{tj}^- \geq 0 \qquad (\forall\, j, t) \qquad (4.4)$$

The objective function (4.1) of the mathematical model developed by Mingers & O'Brien, 1995 aims to minimize the differences between the groups in order to generate an even distribution of the students. With constraint (4.2) the positive and negative deviations from the average values is calculated. Constraint (4.3) ensures that each student is assigned to exactly one group.

Once the example problem is solved using the mathematical model developed by Mingers & O'Brien, 1995, the student groups generated will be as in Figure 4.1. The similarity of assigning a group of students to study groups, a number of workers to some projects, and a number of crew members to flights is clear. Also, it is comprehensible that all these problems have their own sets of constraints. As a result, the equitable partitioning problem and its linear programming model approach will be taken as the solution method of the problem considered in this study.

It takes CPLEX 145.87 seconds to solve optimally on a computer using a 1.8 GHz microprocessor and 8 GB of RAM. The percentage differences and deviations of the groups organized are shown at Table 4.1.

TABLE 4.1
Group Averages and Deviations

Group Averages and Deviations

Group	Maths		Science		Language		History		Arts	
	Average	% Deviation	Average	% Deviation	Average	% Deviation	Average	% Deviation	Average	% Deviation
1	80.6	2.15	73.8	2.5	70.2	0.14	69.4	2.97	76	1.2
2	78	1.14	72.2	0.28	69.6	0.71	65.4	2.97	76.4	1.73
3	78.6	0.38	70.6	1.94	70	0.14	64.6	4.15	74.2	1.2
4	78.4	0.63	71.4	0.83	70.4	0.43	70.2	4.15	73.8	1.73

4.5 A MIXED-INTEGER LINEAR PROGRAMMING MODEL FOR THE SOLUTION OF THE CREW ASSIGNMENT PROBLEM

In order to model and solve the crew scheduling problem considered in this paper, the problem is first defined, then the linear mathematical model is developed transforming the model presented by Mingers and O'Brien (1995) to the problem defined. Then, a series of cases are generated based on the problem definition and these cases are solved using the developed model. The results are analyzed.

4.5.1 PROBLEM DEFINITION

The crew scheduling problem considered in this paper consists of assigning a number of (i) crew members with various attributes (j) such as gender, language competency, experience, first-class working eligibility, and business-class working eligibility to flights the required number of crew members is known, besides the required cabin classes is also known in a way that the assignments are made as evenly as possible. In other words, the main objective of this problem is to satisfy the required number of cabin crew assignments to the flights while providing the cabin class requirements for each of the flights in a way that for each of the attributes taken into consideration, the assignments will be made evenly. For example, if a flight requires six cabin crew, with one for business class, then the crew members assigned to this flight will not comprise of fully male or female, and the average language competency and experience of the crew members assigned to this flight will be as close to the general average of all the candidates (other crew members) to provide equality of assignments for all the flights.

To solve this problem, a mixed-integer linear mathematical model is developed transforming the mathematical model proposed by (Mingers and O'Brien, 1995) and presented in the previous chapter.

4.5.2 THE MIXED-INTEGER LINEAR MATHEMATICAL MODEL

Indices:

$$i: \text{Crew Members} \left(i = 1,2,3,...,I\right)$$

$$j: \text{Attributes} \left(j = 1,2,3,...,J\right)$$

$$f: \text{Flights} \left(f = 1,2,3,...,F\right)$$

Parameters:

$$P_{ij} = \text{The attribute } j \text{ value for crew member } i$$

$$B_i = \begin{cases} 1 \text{ if crew member } i \text{ has the eligibility to be assigned to business class} \\ 0 \text{ otherwise} \end{cases}$$

$$C_i = \begin{cases} 1 \text{ if crew member } i \text{ has the eligibility to be assigned to first class} \\ 0 \text{ otherwise} \end{cases}$$

N_f = The required nuber of crew members to be assigned to flight f

$$\theta_f = \begin{cases} 1 \text{ if flight } f \text{ requires at least one crew for business class} \\ 0 \text{ otherwise} \end{cases}$$

$$\mu_f = \begin{cases} 1 \text{ if flight } f \text{ requires at least one crew for first class} \\ 0 \text{ otherwise} \end{cases}$$

Decision Variables:

$$X_{if} = \begin{cases} 1 \text{ if crew member } i \text{ is assigned to flight } f \\ 0 \text{ otherwise} \end{cases}$$

d_{fj}^- = negative deviation of the crew members assigned to flight f for attribute j

d_{fj}^+ = positive deviation of the crew members assigned to flight f for attribute j

Objective Function:

$$\min Z = \sum_f \sum_j d_{fj}^- + d_{fj}^+ \tag{4.5}$$

s.t.

$$\sum_f X_{if} \leq 1 \forall i \tag{4.6}$$

$$\sum_i X_{if} = N_f \forall f \tag{4.7}$$

$$\sum_f X_{if} \times P_{ij} - d_{fj}^+ + d_{fj}^- = \frac{\sum_i P_{ij} \times N_f}{I} \forall f, j \tag{4.8}$$

$$\sum_i X_{if} \times B_i \geq \theta_f \forall f \tag{4.9}$$

$$\sum_i X_{if} \times C_i \geq \mu_f \forall f \qquad (4.10)$$

$$X_{if} \in \{0,1\} \forall i, f$$

$$d_{fj}^-, d_{fj}^+ \geq 0 \forall f, j$$

The objective function (4.5) of the model aims to minimize the positive and the negative deviations for the crew members assigned to any flight f for any attribute j to maintain an evenly distribution of the crew members amongst the flights. Constraint (4.6) provides that a crew member is assigned to at most one flight. Here, the total number of crew members may be more than the total number of crew members required for the flights, as a result of this the sign is $\leq$ instead of $=$. With constraint (4.7) the assignment of the required number of crew members to every flight is guaranteed. Constraint (4.8) calculates the positive and the negative deviation of the crew members assigned to flight f for attribute j. Since the number of crew members required for each flight differs due to security issues and aircraft type, the RHS of the equation is different from the model proposed by Mingers and O'Brien (1995). Constraints (4.9) and (4.10) ensure that at least one crew member who is eligible to work in business class and first class is assigned to the flight if required, respectively.

4.5.3 CASE STUDY

In order to investigate the validity and the efficiency of the mixed-integer linear programming model proposed in this study for the crew scheduling problem in the aviation industry, firstly, a small-size case is generated randomly for 30 crews and 5 flights requiring a total of 18 crew members. For each of the crew members, three attributes are determined randomly in a range. These are gender (F/M), language competency (40–100), and experience (0–30). Also, the eligibility of each of the crew members to be assigned to business class or first class is generated as a binary parameter. Secondly, the required number of crew members to be assigned to each flight is determined randomly between 3 and 8. Also, the requirement for at least one business-class or first-class crew is determined as a binary parameter. The summarized data for the parameters of Case 1 is given in Table 4.2.

Once the data set is generated, the mixed-integer linear programming model proposed here is coded using GAMS. The coded model is solved to optimality using CPLEX solver to find out the solution time for this small-size instance. It is found out that the solution time for this problem is 40,144.62 seconds with an objective function value of 6.13. The optimal assignment of the crew to the flights is given in a summarized manner in Table 4.3. Since this solution time cannot be accepted as a reasonable time for an operational or a tactical level decision, another method is chosen. The problem is solved with a time limit of 120 s., 300 s., 600 s., and 900 s. respectively, and the best solution found in this period is recorded. The objective function values for different runs are given in Table 4.4.

TABLE 4.2
Summarized Data for Case 1

Crew Members			Flights	
Number of Crew Members (I)		30	Number of Flights(F)	5
Gender	Female	16	Total Number of Required Crew	18
	Male	14	Total number of Required First Class	1
Language Competency	Average	72.47	Total number of Required Business Class	3
	Minimum	43		
	Maximum	99		
Experience	Average	13.73		
	Minimum	0		
	Maximum	29		
First Class Eligibility	Eligible	19		
	Not Eligible	11		
Business Class Eligibility	Eligible	18		
	Not Eligible	12		

TABLE 4.3
Summarized Solution for Case 1

Flight	Assigned Crew	Gender	Language Competency	Experience	First Class Eligibility	Business Class Eligibility
1	10	F	51	10	0	0
	21	M	90	29	1	1
	27	M	75	13	1	1
	30	F	73	3	1	0
2	8	M	70	7	1	1
	13	M	78	14	0	0
	17	F	70	20	1	1
3	3	M	50	19	1	0
	15	M	54	12	1	1
	16	F	87	4	0	1
	24	F	99	20	1	1
4	11	M	51	28	0	1
	19	M	98	19	1	0
	22	M	71	8	1	0
	29	F	70	0	1	1
5	18	F	72	12	1	0
	26	M	80	23	0	1
	28	F	64	6	1	0

TABLE 4.4
Objective Function Values for Different Runs

Solution Time	Optimal (40,144 s.)	120 s.	300 s.	600 s.	900 s.
Objective Function	6.13	7.07	8.13	6.93	6.67

As seen in Table 4.4, a peculiarity is noticed for the solution of the problem for 300 s., as the objective function value is found to be 8.13 while it was 7.07 for the 120 s. run. To avoid any error, the model was solved several times, but the same result was found for all runs. As a result, this irregularity is concluded as the working mechanism of the solver CPLEX. Excluding the objective function found for 300 s. run, it may be seen that the objective function value decreased for 600 s., and 900 s. run.

Once the small-size case with 30 candidate crew members and five flights is solved using the proposed model, four additional cases are generated using the same methodology defined. Case 2 is created with 50 crew members and 8 flights, Case 3 with 100 crew members and 10 flights, Case 4 with 150 crews and 25 flights, and finally Case 5 with 350 crews and 50 flights.

Since the time needed to solve small-size case (Case 1) is above expectation, other cases are not solved to optimality but using the same method, each of the cases is solved for 120 s., 300 s., 600 s., 900 s., respectively. The results are shown in Table 4.5.

The results to the additional cases shown in Table 4.5 indicate that even for 120 s. the model can generate a feasible solution to the problem. Again, the peculiarity detected for Case 1 occurred for Case 2 for 600 s. run and 900 s. run, and Case 5 for 900 s. run. which are shown in bold in Table 4.5.

4.6 CONCLUSIONS AND FUTURE REMARKS

The airline industry is a highly competitive industry where the control of cost parameters is crucial. Since, some of the indicators such as crude oil prices, economic downturns, political environment, pandemics, and volatility in customer demand and expectation cannot be controlled by the industry, controlling and cost-effectively managing the rest becomes the game changer for the airlines. One of the most important

TABLE 4.5
Objective Function Values for Different Runs of Additional Cases

Case	Number of Crew (I)	Number of Flights (F)	Objective Function Value for Different Runs			
			120 s.	300 s.	600 s.	900 s.
2	50	8	14.40	12.56	**14**	**21.28**
3	100	10	15.32	13.42	10.46	10.34
4	150	25	149.6	104.44	92.31	75.84
5	350	50	366.65	238.98	162.78	**187.66**

cost issues that require analytical management is the crew or staff cost. As a result of this, the application of smart scheduling techniques for the crew is a rational way of reducing this cost. Besides, such a method may improve customer satisfaction and loyalty from the service quality point of view.

In this work, the cabin crew scheduling or assignment problem is taken into consideration. A mixed-integer linear programming model is developed transforming an existing model in the literature and presenting the problem as an equitable partitioning problem.

The solution performance of the model is investigated by generating five different cases with different sizes and complexity which may easily be transformed into a real-life application with existing data of an airline.

The assignment problem in this work relies on different attributes of individuals, where the objective is to partition these individuals as even as possible. From the airline industry's point of view, this means assigning equal cabin crew teams to flights considering different attributes. The attributes considered here were gender, language competency, and experience. However, the mathematical model proposed here can be converted and tailored for specific objectives if further attributes need to be applied. These attributes may vary from airline to airline, but the goal remains the same.

It is seen that even for a small case with 30 candidate crew members and 5 flights, the solution time to find the optimal value for the objective function is found to be 40,144.62 seconds, which is not acceptable for a tactical or operational-level decision. Hence, multiple metaheuristic algorithms may be used to find solutions more quickly. Also, the model may be improved by adding constraints.

It is clear that the search for better methods to maintain and improve customer satisfaction and profitability for the airline industry is an endless challenge.

REFERENCES

Badanik, B., Duc, M.L., & Kandera, B., "Understanding scheduling preferences of airline crews", *Transportation Research Procedia*, 59, 223–233, (2021).

Deveci, M., & Demirel, N.C., "A survey of the literature on airline crew scheduling", *Engineering Applications of Artificial Intelligence*, 74, 54–69, (2018).

Eltoukhy, A.E., Chan, F.T., & Chung, S.H., "Airline schedule planning: A review and future directions", *Industrial Management & Data Systems*, 117(6), 1201–1243, (2017).

Erdeş, H., "Eşit bölümleme (EP) problemi yaklaşımı ile proje bazlı çalışma kapsamında görev ve bireylerin gruplandırılması", Konya Technical University, master's thesis, (2021).

Fathian, M., Saei-Shahi, M., & Makui, A., "A new optimization model for reliable team formation problem considering experts' collaboration network", *IEEE Transactions on Engineering Management*, 64(4), 586–593, (2017).

Holmberg, K., "Formation of student groups with the help of optimisation", *Journal of the Operational Research Society*, 70(9), 1538–1553, (2019).

Krass, D., & Ovchinnikov, A. "Constrained group balancing: Why does it work.", *European Journal of Operational Research*, 206(1), 144–154, (2010).

Mingers, J., & O'Brien, F.A., "Creating student groups with similar characteristics: A heuristic approach", *Omega*, 23(3), 313–321, (1995).

O'Brien, F., & Mingers, J. "A heuristic algorithm for the equitable partitioning problem", *Omega*, 25(2), 215–223, (1997).

Quesnel, F., Desaulniers, G., & Soumis, F., "A branch-and-price heuristic for the crew pairing problem with language constraints", *European Journal of Operational Research*, 283(3), 1040–1054, (2020).

Rahmanniyay, F., Yu, A.J., & Seif, J., "A multi-objective multi-stage stochastic model for project team formation under uncertainty in time requirements", *Computers & Industrial Engineering*, 132, 153–165, (2019).

Rubin, P.A., & Bai, L. "Forming competitively balanced teams", *IIE Transactions*, 47(6), 620–633, (2015).

van de Water, T., van de Water, H., & Bukman, C., "A balanced team generating model", *European Journal of Operational Research*, 180(2), 885–906, (2007).

Wen, X., Sun, X., Sun, Y., & Yue, X., "Airline crew scheduling: Models and algorithms", *Transportation Research Part E*, 149, 1–18 (2021). https://www.sciencedirect.com/science/article/pii/S1366554521000788?via%3Dihub

Wen, X., Chung, S., Ji, P., & Sheu, J.-B., "Individual scheduling approach for multi-class airline cabin crew with manpower requirement heterogeneity", *Transportation Research Part E*, 163, 1–23 (2022). https://www.sciencedirect.com/science/article/pii/S1366554522001545?via%3Dihub

Yan, S., & Tu, Y.P., "A network model for airline cabin crew scheduling", *European Journal of Operational Research*, 140(3), 531–540, (2002).

Yan, S., Tung, T.T., & Tu, Y.P., "Optimal construction of airline individual crew pairings", *Computers & Operations Research*, 29(4), 341–363, (2002).

Yildiz, B.C., Gzara, F., & Elhedhli, S., "Airline crew pairing with fatigue: Modeling and analysis", *Transportation Research Part C*, 74, 99–112, (2017).

5 The Internet of Things and Cyber-Physical Systems
Aviation Industry Applications

Mehmet Akif Gunduz, Turan Paksoy, and Engin Hasan Çopur
Necmettin Erbakan University, Konya, Turkey

5.1 INTRODUCTION

Industry 4.0 technologies, such as the Internet of Things (IoT) and Cyber-Physical Systems (CPS), have revolutionized the aviation industry. IoT in the aviation industry enables the collection of vast amounts of data from various sources, such as aircraft systems, weather sensors, and passenger feedback, which can be analyzed to improve service quality, safety, and efficiency. For instance, airlines can use IoT to monitor the performance of their aircraft engines in real-time, predict maintenance needs, and reduce downtime. In the contemporary landscape, Industry 4.0 stands as a revolutionary force, permeating diverse sectors and reshaping the very fabric of our technological ecosystem. The aviation industry is at the forefront of this paradigm shift, a cornerstone of economic, political, and military considerations. The relentless evolution of technology has not only propelled Industry 4.0 into the heart of aviation, but has also become a catalyst for sustained growth and innovation.

This chapter delves into the intricate interplay between Industry 4.0 technologies, focusing on the aviation industry. From low-tech manufacturing to advanced service sectors, the influence of this fourth industrial revolution is far-reaching. Here, we explore the convergence of the IoT and CPS within aviation, elucidating their applications and implications.

With its economic, political, and military ramifications, the aviation industry is strategically positioned in the global landscape. Understanding the transformative impact of Industry 4.0 technologies within this sector is not merely an academic pursuit, but a pragmatic necessity also. The combination of IoT and CPS in aviation promises to enhance operational efficiency and ensure safety, sustainability, and customer satisfaction. The primary objective of this study is to dissect and illuminate the role of IoT and CPS in shaping the trajectory of the aviation industry under the umbrella of Industry 4.0. By exploring the applications of these technologies, we aim

DOI: 10.1201/9781003389187-5

69

to unravel their tangible benefits, potential challenges, and the overarching impact on aviation operations, safety protocols, and overall efficiency.

This chapter is structured to provide a comprehensive journey through the key facets of our exploration. We commence with an overview of Industry 4.0, tracing its historical roots and evolution. From there, we delve into the essential tools driving Industry 4.0, setting the stage for a detailed examination of IoT and CPS. Subsequently, we pivot to the heart of our discussion—applications of IoT and CPS in the aviation industry. Real-world examples, such as the transformative IoT-based baggage handling system, will serve as touchstones to illustrate the tangible impact of these technologies. Finally, we conclude with a reflection on the implications for aviation, emphasizing the need for continued research and development in this dynamic intersection of technology and air travel. Through this organized structure, we aim to provide readers with a nuanced understanding of the present and future of Industry 4.0 in the aviation sector.

5.2 THE RISE OF INDUSTRY 4.0: IOT AND CPS

The term "Industry 4.0" appeared for the first time in the most significant industrial fair in the world, Hannover Messe, in Germany, in 2011. In the opening ceremony, Prof. Wolfgang Wahlster, Director and CEO of the German Research Center for Artificial Intelligence, delivered a keynote speech on the critical points companies must consider for maintaining long-term business success in a competitive market. In his speech, he pointed out that companies keen to have a significant share of the global markets and sustainable economic performance must implement Industry 4.0 technologies into their business activities (Lydon, 2014; Szajna et al., 2020). At the end of the opening ceremony, the audience witnessed the start of the recent Industrial Revolution era, namely the Era of Industry 4.0. Therefore, this section provides the readers with a brief history of Industry 4.0 and a brief overview of Industry 4.0 technologies, explicitly targeting the IoT and CPS.

5.2.1 HISTORY OF INDUSTRY 4.0 REVOLUTION

Humanity had been manufacturing equipment by hand and powered by humans or animals until steam was discovered as an alternative power source in the middle of the 18th century. Steam power enabled the transition from hand production methods to a new manufacturing process in which steam-powered machines emerged as a critical instrument in the manufacturing industry. Therefore, traditional methods, workshops, and business life gradually disappeared. Consequently, a new world was born and shortly occupied by buildings enormously more significant than workshops, known as factories. This genuinely radical transformation was the first step towards industrialization, which reshaped social structure by creating a new social class known as the middle class, while stoking rapid economic growth. When the 20th century dawned, electricity began to be used in mass production, and this was accepted as the birth of the 2nd Industrial Revolution. The 3rd Industrial Revolution was the period that started with the entry of computers into factories in the third quarter of the 20th century. It refers to integrating mechanical technology with electronics, robotics, and IT in production processes. Technologies starting with the letter "C", such as CAD and CAM, that allow computer-controlled applications in the production environment, became widespread in this period. The 4th Industrial Revolution is based on the 3rd Industrial Revolution, and the spectacular advances in digital technology led to the 4th (Romanovs et al., 2019). The 4th Industrial Revolution greatly benefits all industry and service sectors worldwide. It provides valuable tools to collect real-time data, make sense of it, and use it in predictive analysis. These tools include ten advanced technologies that will be briefly explained in the following section.

5.2.2 TOOLS OF INDUSTRY 4.0

Industry 4.0 focuses on improving the efficiency and quality of the operations of the service and manufacturing sectors. Therefore, it offers valuable solutions for complex, capacity and flexibility problems (Rennung et al., 2016). Integrating state-of-the-art technologies into business systems produces these practical and ingenious solutions. These advanced technologies are known as Industry 4.0 technologies. However, there has yet to be a consensus about the indispensable technologies for a company aiming to become a key player in the Industry 4.0 era. To date, several studies have attempted to identify the tools of Industry 4.0 (Elafri et al., 2022). Silvestri et al. (2022) introduced IoT, CPS, and Big Data as the essential tools for implementing lean production in the Industry 4.0 era. In addition, autonomous robots, augmented reality, and additive manufacturing were also considered as other tools even though have less impact on the success of lean production. In a review analysis, 15 different technologies were examined to reveal their relationship with lean management. These technologies include Big Data, Sensors, CPS, Augmented and virtual reality, 3D printing, IoT, Internet of Service, Robotics, Simulation, 3D model and Optimization Algorithms, Machine-to-Machine (M2M), and Data analytics. Industry 4.0 technologies can also be applied to the transition to the digital economy. The leading technologies enabling this transition were given in the national programme "Digital Economy of the Russian Federation" as Big Data, Neural network and AI, distributed ledger systems (Blockchain), quantum technologies, new production technologies, Industrial Internet, components of robotics and sensors, wireless communication technology, virtual and augmented reality technology (Fedotova et al., 2020). An alternative classification was made that manufacturers can use as a helpful guide in the digital transition phase (Ghobakhloo, 2018). According to these classifications, some technologies, such as IoT and CPS, are essential to the Industry 4.0 revolution.

5.2.3 IoT AND CPS

IoT and CPS are related concepts that involve integrating physical and digital systems. CPS and IoT share some common characteristics but differ in scope and focus.

The usage of electronic devices and digital platforms is rapidly increasing in every business. This high usage enables access to the digital world. The transition from the physical to the digital world is based on IoT technologies that facilitate direct communication between these two worlds. In the literature (Falanga et al., 2017), there is no unique definition of what is technically meant by IoT (Shaikh et al., 2017; Yaqoob et al., 2017; Zhang et al., 2017). Atzori et al. (2010) define IoT as "a world-wide network of interconnected objects uniquely addressable, based on standard communication protocols". An alternative definition is introduced by Vermesan et al. (2022) as "dynamic global network infrastructure with self-configuring capabilities based on standard and interoperable communication protocols". The practical implementation of IoT can be achieved by integrating electronic devices such as sensors, actuators, and machines into a business environment to collect, analyze, and finally use data to make accurate decisions (Lampropoulos et al., 2019). This communication network improves the quality of the provided service, reduces energy consumption, utilizes the current resources effectively, and predicts possible or unexpected failures in advance (Aazam et al., 2018).

A CPS system integrates computation, communication, and control processes to enable interaction between physical and cyber worlds. It is characterized by the seamless integration of physical and cyber components, which interact to achieve common goals.

A CPS's core is the ability to sense and respond to changes in the physical environment. The physical components of a CPS may include sensors, actuators, and other hardware devices responsible for collecting data and controlling physical processes. The cyber components of a CPS include software, algorithms, and other digital systems that process data, make decisions, and communicate with other systems.

CPS is used in various applications, such as manufacturing, transportation, healthcare, energy, and building automation. They enable efficient and effective control and automation of complex systems while providing real-time data for monitoring and decision-making.

Leng et al. (2019) propose a digital twin-driven manufacturing CPS that enables parallel control of an intelligent workshop. Biesinger et al. (2019) highlight the importance of data integration in CPSs for effective planning. Ma et al. (2019) present an energy-CPS-enabled management system for energy-intensive manufacturing industries to promote cleaner production. Wang et al. (2019) propose an occupancy-linked energy CPS that uses WiFi probes for occupancy detection to reduce energy waste and improve thermal and ventilation services. Kirchhof et al. (2020) present a model-driven method to describe the software of CPS, their digital twin information system, and their integration. Peng et al. (2020) propose a distributed control framework with a finite-time approach for robustness in CPSs with interdependent physical and computational resources. Nateghi et al. (2020) study finite-time convergent observation algorithms for cyber-attacks and fault reconstruction in electric power networks.

5.3 CASE STUDY: IOT AND CPS APPLICATION IN THE AVIATION INDUSTRY

One of the challenges faced by the aviation industry is the efficient and reliable handling of baggage. According to the SITA Baggage Report 2020, 4.65 billion bags were checked in by passengers in 2019, and 24.8 million bags were mishandled, resulting in an estimated cost of $2.5 billion for the industry (Muskan, 2021). Moreover, mishandled baggage causes inconvenience and dissatisfaction for passengers, as well as operational delays and disruptions for airlines and airports. Therefore, there is a need for a smart baggage handling system that can leverage IoT and CPS technologies to improve the tracking, identification, and routing of baggage throughout the journey.

The IoT-based baggage handling system is an example of how IoT and CPS can be applied in the aviation industry to create a smart solution that improves service quality, safety, and profitability (Rahman et al., 2022). The system demonstrates how IoT devices can collect and transmit data from physical objects, such as bags, to cyber systems, such as databases and applications, where they can be processed and analyzed to provide helpful information and insights. The system also shows how CPS algorithms can control and coordinate physical processes, such as baggage handling, based on cyber inputs and outputs, such as RFID tags and readers. The system illustrates how IoT and CPS can seamlessly integrate physical and cyber worlds in a complex and dynamic environment.

A real-life example of the IoT-based baggage handling system is that implemented by an airline company in 2016. The airline company introduced RFID tags for all checked bags, which can be scanned by RFID readers at various points along the baggage handling process. The RFID tags contain information about the bag's owner, flight number, destination, and priority level. The RFID readers transmit this data to a central database, where the company's mobile app or website can access it. Passengers can use these platforms to track their bags in real-time, from check-in to arrival. They can also receive notifications when their bags are loaded or unloaded from the plane.

The RFID tags also enable a CPS algorithm that optimizes the routing and sorting of the bags according to their destination and priority level. The algorithm uses the data from the RFID tags and readers to determine the best path for each bag through the baggage handling system. The algorithm considers travel time, capacity constraints, connectivity constraints, and priority levels. The algorithm then communicates the optimal routing plan to the RFID readers and IoT devices, such as conveyor belts, scanners, sorters, and loading areas. These devices execute the physical baggage handling processes according to the cyber instructions.

The IoT-based baggage handling system has significantly benefited the airline company and its passengers. According to company website (Muskan, 2021), the system has reduced mishandled bags by 70%, saving millions of dollars in costs and reducing customer dissatisfaction. The system has also increased operational efficiency by reducing manual errors and delays. The system has also enhanced safety by preventing unauthorized access or bag tampering.

5.4 CONCLUSION

In this chapter, we have extensively explored the application of IoT and CPS technologies in the aviation industry, focusing on the smart baggage handling system. While the discussion effectively summarizes key findings, further elucidation of the implications of these findings for the aviation industry would enhance the clarity and applicability of the discussed concepts.

5.4.1 THE IMPACT ON AVIATION OPERATIONS

Integrating IoT and CPS technologies in the aviation industry has transformative implications for daily operations. For instance, the smart baggage handling system implemented by an airline company in 2016 showcases the tangible benefits of these technologies. By leveraging RFID tags and readers, the airline significantly reduced mishandled bags by 70%, resulting in substantial cost savings and reduced customer dissatisfaction. However, it also enhanced operational efficiency by minimizing manual errors and delays. Other aspects of aviation operations adopt similar IoT and CPS applications. Real-time monitoring of aircraft components, weather conditions, and passenger feedback through IoT could enable airlines to predict maintenance needs, optimize flight routes, and enhance in-flight services, contributing to the overall efficiency of aviation operations and elevating the passenger experience.

5.4.2 ENSURING SAFETY IN AVIATION

Safety is paramount in the aviation industry, and IoT and CPS technologies are crucial in fortifying safety measures. Implementing CPS algorithms in baggage handling systems, as illustrated in the airline company example, optimizes the routing and sorting of bags and ensures compliance with security protocols. Imagine a scenario where a baggage handling system identifies a bag deviating from its intended route; the CPS algorithm could trigger immediate alerts, preventing unauthorized access or tampering.

Expanding this concept to broader safety applications, integrating CPS in air traffic management systems could lead to autonomous decision-making capabilities. In the event of potential conflicts or emergencies, a highly autonomous system could reduce air traffic congestion, alleviate the workload on air traffic controllers and pilots, and, most importantly, minimize the risk of accidents.

5.4.3 ENHANCING OVERALL EFFICIENCY

Efficiency is a key metric in any industry, and the aviation sector benefits significantly from the continued integration of IoT and CPS. The ability of these technologies to seamlessly connect the physical and cyber worlds facilitates real-time decision-making and problem-solving. For instance, an IoT-driven predictive maintenance system could detect potential issues in aircraft engines before they escalate, minimizing downtime and ensuring optimal performance.

In the current state of the aviation industry, CPS applications extend beyond baggage handling to optimizing supply chains. IoT sensors in warehouses could communicate with CPS algorithms to streamline inventory management, ensuring that the correct parts are available at the right time. This efficiency level could result in cost savings, reduced environmental impact, and improved overall sustainability in aviation operations.

5.4.4 FUTURE DIRECTIONS AND CONSIDERATIONS

The aviation industry should continue exploring innovative ways to leverage IoT and CPS technologies. Research and development in AI-driven decision-making, enhanced communication protocols, and integration of emerging technologies like blockchain could further elevate the industry's capabilities.

In conclusion, this chapter provides a foundation for understanding the transformative potential of IoT and CPS in the aviation industry. By emphasizing concrete examples and hypothetical scenarios, we aim to inspire further research and development to propel the industry towards increased efficiency, safety, and excellence.

REFERENCES

Aazam, M., Zeadally, S., and Harras, K. A. (2018). Deploying Fog Computing in Industrial Internet of Things and Industry 4.0. *IEEE Transactions on Industrial Informatics*, *14*(10), 4674–4682. https://doi.org/10.1109/TII.2018.2855198

Atzori, L., Iera, A., and Morabito, G. (2010). The Internet of Things: A Survey. *Computer Networks*, *54*(15), 2787–2805. https://doi.org/10.1016/j.comnet.2010.05.010

Biesinger, F., Meike, D., Kraß, B., and Weyrich, M. (2019). A Digital Twin for Production Planning Based on Cyber-Physical Systems: A Case Study for a Cyber-Physical System-Based Creation of a Digital Twin. *Procedia CIRP*, *79*, 355–360. https://doi.org/10.1016/j.procir.2019.02.087

Elafri, N., Tappert, J., Rose, B., and Yassine, M. (2022). Lean 4.0: Synergies between Lean Management Tools and Industry 4.0 Technologies. *IFAC-PapersOnLine*, *55*(10), 2060–2066. https://doi.org/10.1016/j.ifacol.2022.10.011

Falanga, D., Zanchettin, A., Simovic, A., Delmerico, J., and Scaramuzza, D. (2017). Vision-Based Autonomous Quadrotor Landing on a Moving Platform. In *2017 IEEE International Symposium on Safety, Security and Rescue Robotics (SSRR), Safety, Security and Rescue Robotics (SSRR)* (pp. 200–207). IEEE. https://doi.org/10.1109/SSRR.2017.8088164

Fedotova, G. V., Buletova, N. E., Ilysov, R. H., Chugumbaeva, N. N., and Mandrik, N. V. (2020). The Guidelines of Public Regulation in Terms of Digitalization of the Russian Economy with the Industry 4.0 Tools. In E. G. Popkova and B. S. Sergi (Eds.), *Digital Economy: Complexity and Variety vs. Rationality* (pp. 846–855). Springer International Publishing.

Ghobakhloo, M. (2018). The Future of Manufacturing Industry: A Strategic Roadmap toward Industry 4.0. *Journal of Manufacturing Technology Management*, *29*(6), 910–936. https://doi.org/10.1108/JMTM-02-2018-0057

Kirchhof, J. C., Michael, J., Rumpe, B., Varga, S., and Wortmann, A. (2020). Model-Driven Digital Twin Construction. In *Proceedings of the 23rd ACM/IEEE International Conference on Model Driven Engineering Languages and Systems* (pp. 90–101). https://doi.org/10.1145/3365438.3410941

Lampropoulos, G., Siakas, K., and Anastasiadis, T. (2019). Internet of Things in the Context of Industry 4.0: An Overview. *International Journal of Entrepreneurial Knowledge*, *7*(1), 4–19. https://doi.org/10.2478/ijek-2019-0001

Leng, J., Zhang, H., Yan, D., Liu, Q., Chen, X., and Zhang, D. (2019). Digital Twin-Driven Manufacturing Cyber-Physical System for Parallel Controlling of Smart Workshop. *Journal of Ambient Intelligence and Humanized Computing*, *10*(3), 1155–1166. https://doi.org/10.1007/s12652-018-0881-5

Lydon, B. (2014). *The 4th Industrial Revolution, Industry 4.0, Unfolding at Hannover Messe 2014*. https://www.automation.com/en-us/articles/2014-1/the-4th-industrial-revolution-industry-40-unfoldin

Ma, S., Zhang, Y., Lv, J., Yang, H., and Wu, J. (2019). Energy-Cyber-Physical System Enabled Management for Energy-Intensive Manufacturing Industries. *Journal of Cleaner Production*, *226*, 892–903. https://doi.org/10.1016/j.jclepro.2019.04.134

Muskan. (2021). *8 Applications of IoT in Aviation Industry*. https://www.analyticssteps.com/blogs/8-applications-iot-aviation-industry

Nateghi, S., Shtessel, Y., and Edwards, C. (2020). Cyber-Attacks and Faults Reconstruction Using Finite Time Convergent Observation Algorithms: Electric Power Network Application. *Journal of the Franklin Institute, 357*(1), 179–205. https://doi.org/10.1016/j.jfranklin.2019.10.002

Peng, H., Liu, C., Zhao, D., Ye, H., Fang, Z., and Wang, W. (2020). Security Analysis of CPS Systems under Different Swapping Strategies in IoT Environments. *IEEE Access, 8,* 63567–63576. https://doi.org/10.1109/ACCESS.2020.2983335

Rahman, M. S., Manickam, S., and Ul Rehman, S. (2022). Role of Internet of Things in Aviation Industry: Applications, Challenges, and Possible Solutions. In *2022 International Conference on Informatics Electrical and Electronics (ICIEE)*, Yogyakarta, Indonesia (pp. 1–6). https://doi.org/10.1109/ICIEE55596.2022.10010233

Rennung, F., Luminosu, C. T., and Draghici, A. (2016). Service Provision in the Framework of Industry 4.0. *Procedia – Social and Behavioral Sciences, 221,* 372–377. https://doi.org/10.1016/j.sbspro.2016.05.127

Romanovs, A., Pichkalov, I., Sabanovic, E., and Skirelis, J. (2019). Industry 4.0: Methodologies, Tools and Applications. In *2019 Open Conference of Electrical, Electronic and Information Sciences (EStream)* (pp. 1–4). https://doi.org/10.1109/eStream.2019.8732150

Shaikh, F. K., Zeadally, S., and Exposito, E. (2017). Enabling Technologies for Green Internet of Things. *IEEE Systems Journal, 11*(2), 983–994. https://doi.org/10.1109/JSYST.2015.2415194

Silvestri, L., Gallo, T., and Silvestri, C. (2022). Which Tools Are Needed to Implement Lean Production in an Industry 4.0 Environment? A Literature Review. *Procedia Computer Science, 200,* 1766–1777. https://doi.org/10.1016/j.procs.2022.01.377

Szajna, A., Stryjski, R., Wo, W., and Chamier-Gliszczy, N. (2020). Assessment of Augmented Reality in Manual Wiring. *Sensors, 20*(17), 4755.

Vermesan, O., Friess, P., Guillemin, P., Gusmeroli, S., Sundmaeker, H., Bassi, A., Jubert, I. S., Mazura, M., Harrison, M., Eisenhauer, M., and Doody, P. (2022). Internet of Things Strategic Research Roadmap. In O. Vermesan & P. Friess (Eds.), *Internet of Things – Global Technological and Societal Trends from Smart Environments and Spaces to Green Ict* (pp. 9–52). River Publishers. https://doi.org/10.1201/9781003338604-2

Wang, W., Hong, T., Li, N., Wang, R. Q., and Chen, J. (2019). Linking energy-cyber-physical systems with occupancy prediction and interpretation through WiFi probe-based ensemble classification. *Applied Energy, 236,* 55–69. https://doi.org/10.1016/j.apenergy.2018.11.079

Yaqoob, I., Ahmed, E., Hashem, I. A. T., Ahmed, A. I. A., Gani, A., Imran, M., and Guizani, M. (2017). Internet of Things Architecture: Recent Advances, Taxonomy, Requirements, and Open Challenges. *IEEE Wireless Communications, 24*(3), 10–16. https://doi.org/10.1109/MWC.2017.1600421

Zhang, Y., Qian, C., Lv, J., and Liu, Y. (2017). Agent and Cyber-Physical System Based Self-Organizing and Self-Adaptive Intelligent Shopfloor. *IEEE Transactions on Industrial Informatics, 13*(2), 737–747. https://doi.org/10.1109/TII.2016.2618892

6 Analyzing Airport Service Quality through Sentiment Analysis Using Machine Learning Techniques

Batin Latif Aylak
Turkish-German University, Istanbul, Turkey

6.1 INTRODUCTION

Airports are essential elements of the transportation sector, and travellers' pleasure is greatly influenced by the quality of the services they receive. There is a greater demand than ever for effective and high-quality airport services due to the rising number of people flying throughout the world. The number of travellers by air increased from 310 million in 1970 to 4.3 billion in 2018. As a result of its ease and speed of travel, air travel has gained popularity as a means of transportation. It is essential for national economies, tourism, and international trade (THE WORLD BANK, 2018). Therefore, it is crucial to pinpoint the elements that influence the standard of airport services and raise customer satisfaction. Machine learning algorithms have been used to assess the quality of the airport service and provide recommendations for improvement depending on the parameters found. The air transportation network depends on airports, which are increasingly run as multiservice facilities. Airport management firms compete internationally today in a highly competitive climate to draw more travellers and airlines through a variety of services (Gitto & Mancuso, 2017). Sentiment analysis, or analyzing digital text for emotional tone, holds significant practical implications across a wide range of domains and industries.

This study specifically examines four widely used machine learning models: decision tree, random forest, logistic regression and long short-term memory (LSTM). Each of these techniques has been successfully applied within the literature to a multitude of classification problems (Pranckevicius & Marcinkevičius, 2017). These models were used to analyze the airline review dataset, after which their performance was evaluated across a spectrum of metrics, with a particular emphasis on the area under the curve (AUC) score and accuracy (Parker, 2011). Additionally, matrices such as F1 score and precision were considered to measure the performance and efficacy of each model. All measurement criteria used in this study have been used in previous research within the field, providing an excellent comparison tool for the generated models (Jain et al., 2021b).

The performance of each model was analyzed using the same airline review dataset from Kaggle (Bhojani, 2021) to identify the most suitable model for ASQ evaluation in order to provide guidance and recommendations to both practitioners and researchers of sentiment analysis.

After pre-processing, the data were entered into the model. The pre-processing approach employed, such as vectorizing, lemmatizing, deleting stop words from the data, or tokenizing, can significantly affect how well natural language processing techniques function. Therefore, it must be carefully chosen (Angiani et al., 2016). The data used in the LSTM model were tokenized, whereas the data used in the other three models had undergone several pre-processing processes.

This study aims to address the complex task of multiclass sentiment analysis within the context of airport service quality (ASQ) evaluation. The primary objective is to identify the most effective machine learning techniques for accurately discerning sentiment polarity in airline reviews through rigorous comparison and evaluation of pre-existing methodologies.

By examining various machine learning models that have been applied to improve the calibre of airport services, this study adds to the body of literature. To pinpoint the elements that influence traveller satisfaction and raise the standard of airport services, statistical methods such as logistic regression, decision trees, and random forest models have been used. Numerous studies have used decision trees to identify the factors that are most important in determining the standard of airport services and to provide recommendations for improvement based on the identified factors. Recurrent neural networks with the ability to learn long-term dependencies, such as long short-term memory, have been used to forecast the quality of airport services. This study sheds light on how machine learning models can be used to examine the effectiveness of airport services and raise visitor happiness.

The following section of this paper provides a detailed literature review. The methodology is explained in the third section, after which the application and results are presented. Then a few actual case studies are presented. Conclusions and future directions of work are discussed in the final section.

6.2 LITERATURE REVIEW

The quality of airport services is an important aspect of the aviation sector since it has a huge impact on travellers' satisfaction levels, as well as their loyalty to airports. Numerous studies have been conducted on various areas of airport service quality, including the physical qualities of the airport, employee loyalty, and passenger characteristics (Li et al., 2022; Usman et al., 2022).

However, conventional ways of evaluating the quality of the airport service, such as surveys, can be time-consuming and expensive (Li et al., 2022).

Researchers have used sentiment analysis to examine passengers' impressions of the quality of airport service in order to resolve this issue. Based on this information, machine learning algorithms can be trained to categorize the sentiment of the text as either good, negative, or neutral (Bae & Chi, 2021; Li et al., 2022).

Moreover, the another study seeks to evaluate how various machine learning algorithms perform when analyzing sentiment in Twitter data (Poornima & Priya 2020).

Sentiment analysis is a type of natural language processing that can recognize and extract subjective data from text data, including reviews and social media posts. Sentiment analysis has been utilized in several studies to assess the level of airport service. For instance, using sentiment analysis, researchers did a content study of travellers' opinions of the Honolulu International Airport's airport service quality. They discovered that many elements, including airport amenities, employee courtesy, and security protocols, had an impact on travellers' impressions of the calibre of airport services (Bae & Chi, 2021). Similar to this, Bezerra and Gomes (2015) employed sentiment analysis to investigate how passenger attributes and service quality parameters affected their overall happiness with an airport.

Other studies have predicted passenger satisfaction with airport service quality using machine learning approaches. At the Soekarno-Hatta International Airport, Kurniawan et al. (2017) employed a decision tree algorithm to forecast how satisfied travellers were with airport service. They discovered that the most important variables affecting passengers' pleasure were airport amenities, employee service, and security measures.

Airport managers consider airport reviews to be an important indicator of performance. The consumer opinions expressed in these reviews often have a large impact on management and business decisions. Much work has been done to date using machine learning approaches to classify reviews into appropriate categories, including both binary and multiclass classification (Eboli et al., 2022). The existing literature includes studies representing a wide range of measurement matrices, pre-processing steps, data sources, etc. which can aid researchers in developing models for sentiment analysis of review data (Jain et al., 2021a).

Integrating passengers' tweets into an evaluation framework using classification through CNN and LSTM models represents a novel approach to measuring ASQ. This research established a versatile framework for ASQ measurement by leveraging passengers' tweets, regardless of language, allowing the framework to accommodate non-English speaking countries where a variety of languages prevail in passenger communications. This study utilized a substantial dataset, including tweets in both English and Arabic, sourced from four airports. Deep learning models, namely CNN and LSTM, were employed to extract passenger evaluations from tweets, after which the performance of the models was compared and could not be found in the search results. However, there are related papers that discuss the use of deep learning models, including CNN and LSTM, to measure airport service quality using passengers' tweets (Barakat et al., 2021).

The uses, potential, and unresolved issues of deep learning in air traffic management are surveyed in this study. It covers a wide range of subjects, including forecasting flight delays, predicting aircraft trajectories, and forecasting air traffic flow. Additionally, it covers the application of deep learning models for sentiment analysis of passenger tweets on airport services, which can offer information for ASQ evaluation in line with the original query. Additionally, the study emphasizes how crucial it is to train deep learning models in aviation using domain-specific data, which is in line with the empirical results described in the original query (Pinto Neto et al., 2023).

Inconsistent spelling, a lack of capitalization, and the informal grammatical structures of Saudi dialects are only a few of the difficulties in reading Arabic tweets that are highlighted in these publications. Despite these difficulties, academics have developed a number of models and methods for examining Arabic tweets, such as sentiment analysis and categorization algorithms (Albahli, 2022; Alruily, 2021; Alrashidi et al., 2023).

The study uses topic sentiment analysis to examine the public discourse on the Trayvon Martin controversy through social media (Ignatow et al., 2016). Another study examines how content ideology—emotionally charged ideas communicated through sentiments—affects the polarization of opinions on social media (Sun et al., 2023). By concentrating on the measurement and analysis of opinions represented in text data, they investigate the field of sentiment analysis. With a focus on natural language processing (NLP) and machine learning strategies, the paper presumably examines numerous strategies and techniques used to extract feelings from textual content. It discusses the value of sentiment analysis in uses including social media monitoring, market research, and analysis of consumer comments (Bhadane et al., 2015).

Sentiment analysis has also been conducted on larger individual reviews such as blogs reviewing airports, separating the reviews into positive and negative. Accounting for both aviation and non-aviation services, this research shows the possibility of using multiple categorical features integrated into input data to train a machine learning model (Gitto & Mancuso, 2017).

Over the past few decades, a great deal of effort has been paid to measuring and improving consumer satisfaction among airport visitors from various academic fields. Numerous studies have focused on operational effectiveness and productivity measurements in airport environments. However, a thorough evaluation of airport performance necessitates a holistic perspective that goes beyond simple operational considerations. In response, a growing body of research on measuring airport performance has moved its emphasis to models that take customers' demands and opinions of service quality into account.

Using five distinct machine learning algorithms—Logistic Regression, XGBoost, Support Vector Machine, Random Forest, and Naive Bayes—this study examined the attitudes of airline passengers. When compared to other algorithms, the study indicated that XGBoost produced the most accurate findings. The study measured the level of airport service using information from websites like SKYTRAX, Google reviews, and Tripadvisor (Homaid & Moulitsas, 2022).

During the COVID-19 pandemic, this study analyzed customer feedback and measured airport service quality using natural language processing (NLP) and machine learning approaches. According to the study, the exceptional impact of COVID-19 on airport rules and procedures made it essential for airport administrators to monitor the quality of airport services effectively. According to the report, airport administrations require accurate data to understand travellers' expectations for airport services and to assist development initiatives (Li et al., 2022).

The study looked at various machine learning applications in smart airports as well as the usage of deep learning techniques for airport and bus services in the airport zone. The study found that use of machine learning approaches to predict traveller behaviours, streamline procedures, and improve security could improve the quality of airport services (Huang & Zhu, 2021).

In a study that was published in MDPI, the content of 1,341 review comments that were submitted on various platforms were analyzed to determine how travellers felt about the quality of the airport services. The study used machine learning algorithms to analyze the data and discovered that many factors, including service quality dimensions and passenger characteristics, had an impact on travellers' satisfaction with airport service quality (Bae & Chi, 2021).

Social media was employed in another study to perform a sentiment analysis on airport service. To assess it, the researchers examined a dataset of Twitter accounts from London Heathrow Airport. The study discovered that sentiment analysis might be used to gauge the effectiveness of airport services and raise customer happiness (Martin-Domingo et al., 2019).

Another study evaluated the quality of the airport service using sentiment analysis of Google reviews. The study concluded that Google reviews might be used to enhance airport service quality using a complementary approach to quantify perceived service quality (Halpern & Mwesiumo, 2021).

In conclusion, evaluating the quality of airport services by sentiment analysis using machine learning algorithms is a promising strategy. It can offer insightful information about how travellers judge the calibre of airport services and assist airport management in identifying areas for development. To create precise and dependable sentiment analysis models for assessing the calibre of airport services, more study is necessary.

6.3 METHODOLOGY

This study utilized a specialized dataset encompassing reviews for 497 airlines (SKYTRAX, 2023). The dataset was comprised of diverse review types, totalling 23,171 instances, with each instance featuring 17 distinct attributes. The reviews received either a positive label, signifying recommendation, or a negative label, indicating non-recommendation.

To assess the efficacy of different methodologies in understanding and classifying these reviews, the dataset was partitioned into separate subsets for model training and testing model performance, using an 80:20 ratio. Three machine learning models—decision tree, random forest, and logistic regression—and a deep learning model utilizing LSTM were employed as binary classifiers to conduct sentiment analysis of the reviews.

6.3.1 DECISION TREE

Decision tree is a method of supervised learning method that may be applied to both regression and classification problems. Until a halting condition is satisfied, it operates by recursively dividing the data into subsets according to the most important features. The outcome is a structure that resembles a tree, with each leaf node denoting either a numerical value or a class label, and each inside node representing a feature or a decision rule.

Decision trees operate by recursively partitioning datasets into smaller subsets depending on an attribute's value with the objective of building a model that uses input variables to forecast a target variable. Decision trees have previously been used as binary classifiers in sentiment analysis to effectively categorize text as having a positive or negative sentiment (Sharma & Dey, 2012).

6.3.2 RANDOM FOREST

Several decision trees are combined in the Random Forest ensemble learning technique to increase the model's robustness and accuracy. It functions by building a collection of decision trees using features and data subsets that are chosen at random. The average of all the trees in the forest's forecasts yields the outcome.

Random forest models go one step further, building many decision trees during the training process. In classification, the output of the model is the category that is selected by the majority of the individual trees (Al-Amrani et al., 2018). Few studies have examined the quality of airport services using random forest. However, these results imply that random forest can be a beneficial tool for figuring out the variables that affect traveller pleasure and enhancing the calibre of airport services. In general, decision trees have been utilized in numerous studies to pinpoint the most crucial elements influencing the calibre of airport services and offer suggestions for improvement based on the pinpointed elements. Decision trees can assist airport management teams with facility renovations and service quality improvements.

6.3.3 LOGISTIC REGRESSION

Logistic regression is a method of supervised learning technique used for binary classification tasks. It functions by employing a logistic function to predict the probability of the binary outcome. A decision border dividing the two classes is the outcome.

Logistic regression is a statistical technique used for datasets in which one or more independent variables influence an outcome. Variables with only two possible values, also known as dichotomous variables, are utilized to measure the outcome. In the case of sentiment analysis, logistic regression categorizes text as having either a positive or negative sentiment. The logistic regression model operates by making predictions about the likelihood of an event—in this case, the likelihood that a text would be positive or negative—based on the values of independent variables—in this case, the words within the text. The text is subsequently assigned to the category given the highest probability by the model (Ramadhan et al., 2017). Numerous studies have used logistic regression extensively to examine the factors affecting airport service quality and how they affect traveller satisfaction. This approach gives airport management teams helpful insights for raising service standards and improving the overall passenger experience.

6.3.4 LONG SHORT-TERM MEMORY (LSTM)

A particular kind of recurrent neural network called an LSTM is able to identify long-term relationships in sequential input, such as text. It is especially helpful for jobs involving sentiment analysis and natural language processing. A memory cell that can hold information for a long time, and three gates that regulate the flow of information into and out of the cell are the two main components of an LSTM.

It is a type of recurrent neural network that can learn long-term dependencies within sequential data like text. While LSTM can perform better than traditional techniques, it requires more computational complexity and training to do so (Rehman et al., 2019). Time-series data prediction and demand forecasting for air travel can both benefit from using LSTM.

Before data were fed into the machine learning models, a series of pre-processing steps were taken to allow for optimal model training. These steps included eliminating duplicates; imputing missing values; encoding categorical variables; tokenizing textual data; expunging digits, stop words, and special characters; lemmatization; and ultimately, the vectorization of review text data. This comprehensive pre-processing procedure was designed to maximize model performance, considering the idiosyncrasies of the different models under consideration. While some pre-processing steps might inherently favour certain models, the procedures adopted in this study are universally applicable to all models with the exception of deep learning (Angiani et al., 2016). For the deep learning model (LSTM), the input data was only tokenized before training.

Following dataset refinement, each model underwent rigorous training, bolstered by cross-validation strategies. The assessment of model effectiveness went beyond simply accuracy to encompass performance metrics including precision, F1 score, and AUC score, in addition to accuracy. Collectively, these performance metrics offer insight into each model's ability to categorize sentiments accurately and their capacity to be generalized across other fields.

The research methodology was devised to allow for a methodical exploration of multiclass sentiment analysis. By utilizing an array of machine learning models and data pre-processing protocols, the study endeavoured to identify the most applicable technique for classifying latent sentiments embedded within airline reviews.

6.4　APPLICATION AND RESULTS

The performance of the four models differed noticeably from one another, according to a study of the research findings. AUC (Area Under the Curve), F1, accuracy, and precision were the four performance metrics used in the comparative evaluation of model performance; the specifics are given below. The metrics that are being presented cover all the crucial areas of machine learning model performance evaluation.

An indicator of a binary classifier's ability to distinguish between positive and negative classes is its AUC. Plotting the genuine positive rate versus the false positive rate at various threshold values yields the result. The final AUC score has a range of 0–1, with 1 representing the best performance at differentiating between feelings.

To evaluate the overall accuracy of a model, the F1 score computes the harmonic average of recall and precision. Higher scores on the F1 scale, which likewise ranges from 0 to 1, indicate greater accuracy. Following is the F1 formula:

$$F1\_score = \frac{2 \times (\text{precision} * \text{recall})}{(\text{precision} + \text{recall})} \tag{6.1}$$

where

$$\text{precision} = \frac{\text{true positives}}{\text{true positives} + \text{false positives}} \tag{6.2}$$

and

$$\text{recall} = \frac{\text{true positives}}{\text{true positives} + \text{false negatives}} \tag{6.3}$$

By measuring the proportion of examples that were properly predicted to all instances, accuracy assesses how well a binary classifier can predict positive and negative classes. It is calculated by dividing the total number of forecasts by the number of accurate guesses (true positives and true negatives), with a greater value signifying more accurate predictions.

$$\text{accuracy} = \frac{\text{true positives} + \text{true negatives}}{\text{true positives} + \text{false positives} + \text{true negatives} + \text{false negatives}} \tag{6.4}$$

How many of the positive cases that were predicted were actually positives is a measure of precision. It is calculated by dividing the total of genuine positives and false positives by the number of true positives. Fewer false positives are indicated by a higher value. The formula for accuracy is:

$$\text{precision} = \frac{\text{true positives}}{\text{true positives} + \text{false positives}} \tag{6.5}$$

The results of the performance measures for each of the four models are shown in Table 6.1. A number that is closer to 1 indicates a better result for all parameters.

TABLE 6.1
Results of the Models

Model	AUC Score	F1 Score	Accuracy	Precision
Decision Tree	0.9480	0.9306	0.9531	0.9291
Logistic Regression	0.9489	0.9326	0.9546	0.9338
Random Forest	0.9445	0.9312	0.9544	0.9488
LSTM	0.8584	0.8132	0.8740	0.8164

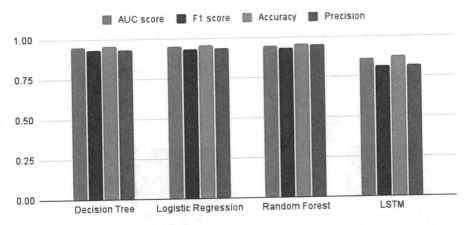

FIGURE 6.1   Performance measurements of different machine learning models.

A graphic representation of the results of the performance measure is shown in Figure 6.1.

As seen in Figure 6.1, logistic regression performed the best on average across all performance measures, though both the decision tree and random forest models performed nearly as well. LSTM had the lowest performance metric scores of the four models, but they were still relatively high, indicating decent performance.

The confusion matrix for each model is shown in Figure 6.2. It displays both accurate and inaccurate forecasts. True categories are on the y-axis, whereas model predictions are on the x-axis. Labels along the x-axis represent the expected categorization, while those on the y-axis represent the actual categorization. True negatives, or instances where the actual category and anticipated category were both negative, are represented in the upper left quadrant. False positives, or instances when the actual category was negative but the model projected it to be positive, are displayed in the upper right quadrant. False negatives are shown in the bottom left quadrant, and true positives are in the bottom right.

It is clear from a detailed analysis of the data that the logistic regression model regularly performs better than other models in terms of many different metrics and criteria. It stands out among the tested models for having the best accuracy, AUC score, and F1 score. This demonstrates how adept logistic regression is at properly classifying emotions while reducing false positive errors.

It is noticeable that the LSTM model performs less well than the other models in comparison. But it is important to consider the environment in which this model was assessed. The LSTM model, in contrast to the other models, was applied to text data that had merely undergone tokenization as pre-processing. It is probable that this less thorough pre-processing strategy played a role in the performance drop.

The study assessed how well various machine learning models performed in assessing airport service quality and raising customer satisfaction using a dataset of airline evaluations. Four machine learning models—decision trees, random forests, logistic regression, and long short-term memory (LSTM)—were used in the study to

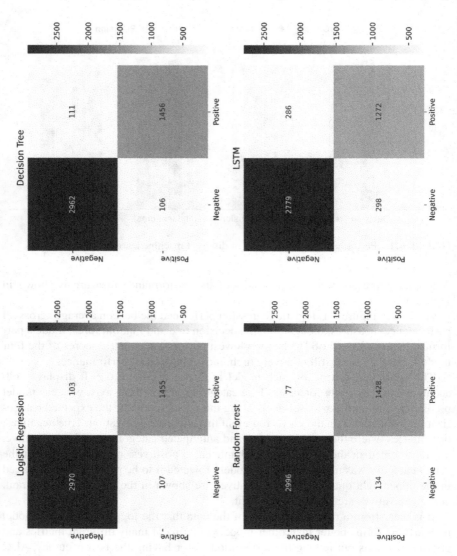

FIGURE 6.2   Confusion matrix for each model.

evaluate the dataset and determine the variables influencing airport service quality. Pre-processing procedures for the data included removing duplicates, imputing missing values, encoding categorical variables, tokenizing textual data, removing special characters, stop words, and digits, lemmatization, and vectorization of review text data. Each model's performance was assessed using metrics such as accuracy, precision, F1 score, and AUC. The decision tree model received the best AUC score, while the random forest model did better than the other models in terms of accuracy and F1 score. The study also determined the most important variables that impact the quality of airport services, such as check-in processes, luggage handling, and aircraft delays.

6.5 CASE STUDY: SENTIMENT ANALYSIS OF AIRPORT SERVICE QUALITY IN THE AVIATION INDUSTRY

Sentiment analysis is an effective approach for assessing customer attitudes and perceptions of airport service quality (ASQ). Since it offers useful information for enhancing airport services and boosting consumer loyalty, sentiment analysis has grown in popularity within the aviation sector in recent years.

The researchers identify satisfiers, dissatisfiers, and performance factors that affect passengers' experiences at the Honolulu International Airport and the world's top airports (Singapore Changi Airport, Haneda Airport, Incheon International Airport, Hamad International Airport, and Hong Kong International Airport) using the 1,341 review comments posted on the SKYTRAX website. The findings indicate that Honolulu International Airport needs to increase facility cleanliness, signage legibility, and personnel civility. According to a context-specific analysis, the most often used words by irate passengers are security, check, flight, queue, and personnel. The most frequently used adjectives when describing a positive airport experience are staff, terminal, clean, time, immigration, and free. These studies offer recommendations and consequences for customers' viewpoints, which may assist airport administrators in improving the quality of airport services. These findings offer recommendations and consequences regarding the viewpoints of consumers, which may assist airport administrators in raising the standard of airport services and updating airport infrastructure (Bae & Chi, 2021).

The dataset from the Twitter accounts of London Heathrow Airport is used in this study to apply the sentiment analysis (SA) method to measure ASQ. The purpose of this study is to investigate how SA techniques can locate novel insights that go beyond what is offered by more conventional approaches. A total of 4,392 tweets are included in the dataset, and the SA identifies 23 qualities that may be compared to other ASQ scales. Findings show that the frequency of passenger referrals to the scale's qualities varies greatly in some instances, and that identifying these discrepancies might give airport management useful information for enhancing the quality of their services. Using sentiment analysis methods applied to blogs, the study evaluates the perceived level of services within the airport sector. The study focuses on the Twitter account dataset for London Heathrow Airport and seeks to quantify ASQ using sentiment analysis. The study clarifies the viability and efficacy of using unstructured textual data to determine passengers' attitudes and views (Martin-Domingo et al., 2019).

For travellers, the standard of airport service is crucial. This report outlines two phases that will help it achieve its goal. In order to determine what is lacking, what needs to be addressed, and what surveys should be conducted, various passengers were first questioned, with an emphasis on dissatisfied passengers. The interview data was coded utilizing the grounded theory method. The findings from this phase showed that key parameters influencing traveller satisfaction include the characteristics of passengers, such as travel preferences and information about services in various airport components. Second, a survey was created to ensure the validity of our findings. Structural equation modelling (SEM) was used with 860 passengers to examine the survey data. The SEM reports highlighted the need for improved accessibility features for travellers with disabilities. Structural equation modelling (SEM) was used with 860 passengers to examine the survey data. The SEM results indicated that the top priority for enhancing the quality of service at airports are to increase the number of facilities for passengers with impairments or illnesses, facilities for packing and wrapping passengers' freight, and information services (Nadimi et al., 2023).

6.6 CONCLUSION

This study into sentiment analysis as a means of assessing airport service quality through machine learning techniques has yielded valuable insights, underscoring the significance of model choice in achieving accurate sentiment classification.

Performance analysis revealed that the logistic regression model consistently outperformed the other models across multiple performance criteria. With the highest accuracy, AUC score, and F1 score, it is a robust contender for sentiment analysis within the context of airport service quality evaluation. The model's ability to accurately categorize sentiments while minimizing false positive errors highlights its efficacy.

The comparatively reduced performance of the LSTM model requires contextual consideration. Its diminished performance can likely be attributed to the less comprehensive pre-processing applied to the text data. This difference in performance underscores the importance of data pre-processing in enhancing model performance, particularly in the domain of deep learning models.

From a practical standpoint, the findings provide actionable implications for airport management and sentiment analysis practitioners. By employing logistic regression models, airport authorities can efficiently gauge passenger sentiments and perceptions, enabling targeted enhancements of service quality. The proficiency of the model in sentiment classification can guide resource allocation and decision-making processes, leading to more informed and effective strategies that improve passenger satisfaction. While this study focused on sentiment analysis related to airport service quality, the methodology and insights can be extrapolated to other industries and domains reliant on customer feedback for decision-making.

Despite having a specific focus on airline service quality, this study's research methodology and findings may be applied to other companies that rely on customer input, offering a guide for future sentiment analysis applications. Other textual data, including blog posts, magazine articles, and social media posts, can also be classified using the methods.

There are several restrictions on this research. The length of each model's training or the potential impact of data volume on model performance were not considered in this investigation. The difficulty of developing the model or the pre-processing pipeline employed for this scenario has not also been discussed. Complexity is a significant factor for many use cases and applications. Less sophisticated models and pre-processing pipelines need to be developed, and more research is needed to determine the implications of training duration and data volume.

This study produced a thorough framework for evaluating service quality in sectors dependent on user-generated evaluations in addition to adding to the empirical understanding of sentiment analysis techniques. In order to present a more comprehensive picture of model success, future research projects should take into account variables other than simple performance measurements, such as model training time, data amount, model complexity, and pre-processing subtleties.

REFERENCES

Al-Amrani, Y., Lazaar, M., & Kadiri, K. E. E. (2018). Random Forest and Support Vector Machine based Hybrid Approach to Sentiment Analysis. *Procedia Computer Science*, *127*, 511–520. https://doi.org/10.1016/j.procs.2018.01.150

Albahli, S. (2022). Twitter Sentiment Analysis: An Arabic Text Mining Approach Based on COVID-19. *Frontiers in Public Health*, *10*, 966779.

Alrashidi, B., Jamal, A., & Alkhathlan, A. (2023). Abusive Content Detection in Arabic Tweets Using Multi-Task Learning and Transformer-Based Models. *Applied Sciences*, *13*(10), 5825.

Alruily, M. (2021). Classification of Arabic Tweets: A Review. *Electronics*, *10*(10), 1143.

Angiani, G., Ferrari, L., Fontanini, T., Fornacciari, P., Iotti, E., Magliani, F., & Manicardi, S. (2016, September). A Comparison between Preprocessing Techniques for Sentiment Analysis in Twitter. In KDWeb, 7(2), 37-56.

Bae, W., & Chi, J. (2021). Content Analysis of Passengers' Perceptions of Airport Service Quality: The Case of Honolulu International Airport. *Journal of Risk and Financial Management*, *15*(1), 5, 1-19

Barakat, H. M., Yeniterzi, R., & Martin-Domingo, L. (2021). Applying Deep Learning Models to Twitter Data to Detect Airport Service Quality. *Journal of Air Transport Management*, *91*, 102003. https://doi.org/10.1016/j.jairtraman.2020.102003

Bezerra, G. C. L., & Gomes, C. F. (2015). The Effects of Service Quality Dimensions and Passenger Characteristics on Passenger's Overall Satisfaction with an Airport. *Journal of Air Transport Management*, *44*, 77–81.

Bhadane, C., Dalal, H., & Doshi, H. (2015). Sentiment Analysis: Measuring Opinions. *Procedia Computer Science*, *45*, 808–814.

Bhojani, J. (2021). Airline Reviews (Version 1) [Data Set]. Kaggle. https://www.kaggle.com/datasets/juhibhojani/airline-reviews

Eboli, L., Bellizzi, M. G., & Mazzulla, G. (2022). A Literature Review of Studies Analysing Air Transport Service Quality from the Passengers' Point of View. *Promet-Traffic & Transportation*, *34*(2), 253–269.

Gitto, S., & Mancuso, P. (2017). Improving Airport Services Using Sentiment Analysis of the Websites. *Tourism Management Perspectives*, *22*, 132–136. https://doi.org/10.1016/j.tmp.2017.03.008

Halpern, N., & Mwesiumo, D. (2021). Airport Service Quality and Passenger Satisfaction: The Impact of Service Failure on the Likelihood of Promoting an Airport Online. *Research in Transportation Business & Management*, *41*, 100667.

Homaid, M. S., & Moulitsas, I. (2022, October). Measuring Airport Service Quality Using Machine Learning Algorithms. In *Proceedings of the 6th International Conference on Advances in Artificial Intelligence* (pp. 8–14).

Huang, H., & Zhu, J. (2021). A Short Review of the Application of Machine Learning Methods in Smart Airports. In *Journal of Physics: Conference Series* (Vol. 1769, No. 1, p. 012010). IOP Publishing.

Ignatow, G., Evangelopoulos, N. and Zougris, K. (2016), Sentiment Analysis of Polarizing Topics in Social Media: News Site Readers' Comments on the Trayvon Martin Controversy. In *Communication and Information Technologies Annual (Studies in Media and Communications)*, Vol. 11, Emerald Group Publishing Limited, Leeds, pp. 259–284. https://doi.org/10.1108/S2050-206020160000011021

Jain, P. K., Pamula, R., & Srivastava, G. (2021a). A Systematic Literature Review on Machine Learning Applications for Consumer Sentiment Analysis Using Online Reviews. *Computer Science Review*, *41*, 100413. https://doi.org/10.1016/j.cosrev.2021.100413

Jain, P., Saravanan, V., & Pamula, R. (2021b). A Hybrid CNN-LSTM: A Deep Learning Approach for Consumer Sentiment Analysis Using Qualitative User-Generated Contents. *ACM Transactions on Asian and Low-Resource Language Information Processing*, *20*(5), 1–15. https://doi.org/10.1145/3457206

Kurniawan, R., Sebhatu, S. P., & Davoudi, S. (2017, January). Passengers' Perspective toward Airport Service Quality (ASQ) (Case study at Soekarno-Hatta International Airport). In *Civ. Eng. Forum* (Vol. 3, No. 1).

Li, L., Mao, Y., Wang, Y., & Ma, Z. (2022). How Has Airport Service Quality Changed in the Context of COVID-19: A Data-Driven Crowdsourcing Approach Based on Sentiment Analysis. *Journal of Air Transport Management*, *105*, 102298.

Martin-Domingo, L., Martín, J. C., & Mandsberg, G. (2019). Social Media as a Resource for Sentiment Analysis of Airport Service Quality (ASQ). *Journal of Air Transport Management*, *78*, 106–115.

Nadimi, N., Mansourifar, F., Shamsadini Lori, H., & Soltaninejad, M. (2023). How to Outperform Airport Quality of Service: Qualitative and Quantitative Data Analysis Extracted from Airport Passengers Using Grounded Theory (GT) and Structural Equation Modeling (SEM). *Iranian Journal of Science and Technology, Transactions of Civil Engineering*, *48*(1), 483–496.

Parker, C. (2011). An Analysis of Performance Measures for Binary Classifiers. In *2011 IEEE 11th International Conference on Data Mining*, Vancouver, BC, Canada. https://doi.org/10.1109/icdm.2011.21

Pinto Neto, E. C., Baum, D. M., Almeida Jr, J. R. D., Camargo Jr, J. B., & Cugnasca, P. S. (2023). Deep Learning in Air Traffic Management (ATM): A Survey on Applications, Opportunities, and Open Challenges. *Aerospace*, *10*(4), 358.

Poornima, A. S., & Priya, K.S. (2020). A Comparative Sentiment Analysis of Sentence Embedding Using Machine Learning Techniques. In *2020 6th International Conference on Advanced Computing and Communication Systems (ICACCS), Coimbatore, India, 2020*. https://doi.org/10.1109/icaccs48705.2020.9074312

Pranckevicius, T., & Marcinkevičius, V. (2017). Comparison of Naive Bayes, Random Forest, Decision Tree, Support Vector Machines, and Logistic Regression Classifiers for Text Reviews Classification. *Baltic Journal of Modern Computing*, *5*(2), 221. https://doi.org/10.22364/bjmc.2017.5.2.05

Ramadhan, W. P., Novianty, A., & Setianingsih, S. T. M. T. C. (2017). Sentiment Analysis Using Multinomial Logistic Regression. In *2017 International Conference on Control, Electronics, Renewable Energy and Communications (ICCREC), Yogyakarta, Indonesia, 2017* (pp. 46–49). https://doi.org/10.1109/iccerec.2017.8226700

Rehman, A. U., Malik, A. K., Raza, B., & Ali, W. (2019). A Hybrid CNN-LSTM Model for Improving Accuracy of Movie Reviews Sentiment Analysis. *Multimedia Tools and Applications*, *78*(18), 26597–26613. https://doi.org/10.1007/s11042-019-07788-7

Sharma, A., & Dey, S. (2012, October). A Comparative Study of Feature Selection and Machine Learning Techniques for Sentiment Analysis. In *Proceedings of the 2012 ACM Research in Applied Computation Symposium* (pp. 1–7). https://doi.org/10.1145/2401603.2401605

SKYTRAX. (2023). "A–Z Airport Reviews – SKYTRAX". [Online]. Available: https://www.airlinequality.com/review-pages/a-z-airport-reviews/. [Accessed: Aug. 28 2023].

Sun, R., Zhu, H., & Guo, F. (2023). Impact of Content Ideology on Social Media Opinion Polarization: The Moderating Role of Functional Affordances and Symbolic Expressions. *Decision Support Systems*, *164*, 113845.

THE WORLD BANK. (2018). "Air Transport, Passengers Carried | Data." [Online]. Available: https://data.worldbank.org/indicator/IS.AIR.PSGR?end=2018&start=1970. [Accessed: Aug. 27 2023].

Usman, A., Azis, Y., Harsanto, B., & Azis, A. M. (2022). Airport Service Quality Dimension and Measurement: A Systematic Literature Review and Future Research Agenda. *International Journal of Quality & Reliability Management, 39*(10), 2302–2322.

7 Blockchain in Aviation
Applications of Blockchain Technology in Aircraft Maintenance Management

Celal Cakiroglu
Turkish-German University, Istanbul, Turkey

7.1 INTRODUCTION

Blockchain technology has the groundbreaking potential to revolutionize several industries, including aviation. It is gaining more and more attention in the aviation sector for its revolutionary potential in the administration of aircraft maintenance. Efficiency, transparency, security, integrity, streamlined supply chain management, improved safety, and customer satisfaction are some of the primary effects of blockchain technology on aviation. By offering a visible and traceable record of the movement of aircraft parts, equipment, and components, blockchain can increase efficiency. According to design standards established by the original equipment manufacturer (OEM), aviation parts can be divided into various types, each of which has its own specific aircraft spare parts inventory management (ASPM) method (Ho et al., 2021). By enabling smart contracts, blockchain technology can guarantee that the appropriate maintenance procedures are performed for each category. Smart contracts automatically initiate operations depending on established circumstances, ensuring that recoverable components are efficiently recycled, throwaway parts are replaced when necessary, and repairable parts receive the proper fixes (Alharby & van Moorsel, 2017). By keeping a decentralized ledger of maintenance activities, repairs, and inspections, it can improve aviation tracking and maintenance. The effectiveness of maintenance operations is increased thanks to the increased openness, which also makes it easier to comply with legal obligations. Data accuracy and integrity are ensured by blockchain technology, which is essential in the aviation industry. A safe and impenetrable maintenance history can be kept by documenting maintenance activities, inspections, and fixes on the blockchain (Lopes et al., 2021). This enables stakeholders such as airlines, maintenance providers, and regulators to have access to accurate and up-to-date records, simplifying compliance and audits. The aviation industry's supply chain management could be simplified by blockchain technology. By offering a streamlined approach for the prompt distribution of income and removing potential delays brought on by revenue-sharing conflicts, it decreases errors, delays, and disputes. This is advantageous to airlines, aircraft managers, and passengers (Akhmatova et al., 2022). Blockchain technology can improve customer

92

DOI: 10.1201/9781003389187-7

happiness and safety in the aviation industry. For safety and regulatory compliance, accurate and current airplane maintenance records are essential. Airlines, maintenance providers, and regulators may instantly obtain and authenticate the relevant information with the help of blockchain technology, which can give a tamper-resistant, transparent, and easily available history of an aircraft's maintenance. This raises customer and safety confidence (Santonino et al., 2018).

Efficient spare part management is crucial for the safe and reliable operation of aircraft. The integration of blockchain technology can streamline this process by creating a shared and immutable ledger that records every transaction and movement of spare parts. Blockchain technology enhances spare part management in the aviation industry by transparency, traceability, preventing counterfeit parts, efficient inventory management, and enhanced compliance. There is provision for transparency by creating a decentralized ledger that records the origin, maintenance history, and authenticity of each spare part. This ensures that the traceability of spare parts is verifiable throughout their lifecycle. By having a transparent record of spare parts, the risk of counterfeit parts entering the supply chain is minimized, which helps prevent potential safety hazards (Schultz et al., 2020). The supply chain entry of counterfeit parts can be stopped by using blockchain technology in spare part management. The validity of each spare part can be confirmed by storing the serial numbers of spare parts on the blockchain along with transactional information and maintenance logs. This guarantees that only authentic, authorized parts are used to maintain aircraft, improving safety and dependability. By giving real-time access into the availability and location of spare components, blockchain technology can enhance inventory management. Locating and acquiring spare parts is made easier when a common ledger is available to all pertinent stakeholders, such as manufacturers, suppliers, maintenance groups, and airlines. By doing this, downtime is decreased, and aircraft maintenance operations are run more effectively overall (Rolinck et al., 2021). The administration of spare parts is made easier by regulatory compliance thanks to blockchain technology. All transactions and movements of spare parts are precisely documented and cannot be altered because of the immutable nature of the blockchain. As a result, stakeholders may readily obtain and verify the essential information, facilitating audits and regulatory inspections. In conclusion, integrating blockchain technology into the administration of airplane spare parts has many advantages, including increased transparency, traceability, preventing the use of counterfeit parts, effective inventory management, and improved regulatory compliance. The aviation sector may enhance spare part management procedures and guarantee the security and dependability of aircraft operations by utilizing blockchain technology (Ahmed and MacCarthy 2023).

The concept of a "digital twin" involves creating a virtual representation of a physical asset, system, or process. In the aviation industry, a digital twin can represent an entire aircraft or its individual components. Blockchain technology can contribute to the creation and management of digital twins in the aviation industry by providing a secure and decentralized platform, predictive maintenance, and enhanced safety (Mandolla et al., 2019). It provides a secure and decentralized platform for storing and sharing data, which is crucial for creating and managing digital twins. By leveraging blockchain technology, stakeholders can access and share data in real-time, facilitating collaboration and enhancing efficiency. The integration of real-time data from sensors and maintenance records into a digital twin can enable predictive maintenance. Potential issues can be identified and addressed before they lead to operational disruptions, enhancing aircraft safety and reducing maintenance costs. Digital twins can significantly enhance aircraft safety by providing a virtual representation of an aircraft or its components. By simulating and predicting performance, virtual testing can be performed, and scenarios can be tested, improving safety and reliability. Overall, the integration of blockchain technology in the creation and management of digital twins in the aviation industry offers numerous benefits, including enhanced collaboration, efficiency, predictive maintenance, and safety. By leveraging blockchain technology, the aviation industry can optimize digital twin processes and ensure the safety and reliability of aircraft operations (Huang et al., 2020; Li & Mardani, 2023; Sasikumar et al., 2023).

Due to developments in technology and design, the aviation sector has seen a swift rise in the diversity and complexity of aircraft spare components. By offering a uniform platform, improved traceability, effective tracking, and increased compliance, blockchain technology can make it easier to manage this heterogeneous terrain. A standardized platform for storing and obtaining maintenance data may be made available through blockchain technology. This minimizes the possibility of errors and misunderstandings by ensuring that all stakeholders, from manufacturers to maintenance teams, have access to correct and current information. Traceability can be improved by using blockchain technology in the management of aviation spare parts. The validity of each spare part can be confirmed by storing the serial numbers of spare parts on the blockchain along with transactional information and maintenance logs. This guarantees that only authentic, authorized parts are used to maintain aircraft, improving safety and dependability. Blockchain technology can offer a reliable tracking system for spare components for aircraft. Locating and acquiring replacement parts is made easier by developing a decentralized ledger that tracks each transaction and movement of spare components. As a result, downtime is decreased and aircraft maintenance operations are run more effectively overall. The administration of spare parts is made easier by regulatory compliance thanks to blockchain technology. All transactions and movements of spare parts are precisely documented and cannot be altered because of the immutable nature of the blockchain. To that end, stakeholders may readily obtain and verify the essential information, facilitating audits and regulatory inspections. Overall, incorporating blockchain technology into the administration of aircraft spare parts has many advantages, such as a uniform platform for storing and retrieving maintenance data, enhanced traceability, effective tracking, and increased compliance. The aviation sector can improve spare part management procedures and guarantee the security and dependability of aircraft operations by utilizing blockchain technology (Kuhle et al., 2021).

For the benefit of their fleets and customers' safety, airlines are responsible for performing required maintenance. By offering a clear and unchangeable record of all maintenance activities, blockchain technology can help to improve aviation maintenance. To protect the security of their fleets and customers, airlines are fully responsible for doing the required maintenance. Using blockchain technology, it is possible to keep an accurate record of all maintenance tasks. As proof of adherence to maintenance schedules and standards, this record can be used as a reliable source of truth during regulatory audits. The use of blockchain supports compliance with roles and duties in the maintenance process by enforcing accountability for all participants. The administration of airplane maintenance can benefit from an open and cooperative ecosystem created by public blockchain technology. The proper maintenance procedures for each class of aircraft parts may be followed thanks to smart contracts. Anyone can sign up for and participate in a public blockchain without asking for permission, which can promote inclusivity and openness. However, privacy issues must be addressed, particularly when distributing private operational and maintenance information (Hewa et al., 2021).

In conclusion, the use of blockchain technology in airplane maintenance management presents exciting opportunities for resolving issues brought on by the complexity of contemporary aviation systems. Blockchain technology has the potential to revolutionize how the aviation industry manages its assets and operations by improving transparency, security, and efficiency in spare part management, contributing to the development of a digital twin, and streamlining maintenance procedures for different aircraft parts. The advantages it offers in terms of safety, dependability, and cost-effectiveness make it an appealing avenue for future investigation and adoption within the aviation sector, even though its implementation may necessitate overcoming technological and regulatory obstacles.

7.2   A BRIEF HISTORY OF BLOCKCHAIN TECHNOLOGY

A blockchain is a decentralized, distributed, immutable ledger that stores the record of ownership of digital assets (Far et al., 2023). Its roots can be traced back to the concept of distributed ledgers and cryptographic principles. The term "blockchain" was first introduced in 2008 in a whitepaper titled "Bitcoin: A Peer-to-Peer Electronic Cash System." Blockchain technology has immense potential in revolutionizing various industries by providing an immutable digital record of transactions and assets (Ashraf & Heavey, 2023).

In 2009, the first blockchain-based cryptocurrency, Bitcoin, was launched, utilizing a transparent and immutable public ledger for transactions. Its innovative proof-of-work consensus mechanism engaged miners in solving intricate puzzles to validate and add blocks to the chain, bolstering security and transaction history integrity. Bitcoin's success fostered the emergence of diverse cryptocurrencies and blockchain ventures. In 2011, Namecoin introduced decentralized domain names, while Ethereum, introduced in 2013, enabled programmable smart contracts and decentralized applications. These self-executing contracts, a hallmark of blockchain progress, automated processes and elevated trust across industries, with Ethereum leading the way. The concept of DApps also gained traction. These decentralized applications aimed to provide alternatives to traditional centralized services, ranging from finance and supply chain management to gaming and social networking. Their appeal lay in enhanced security, transparency, and user control (Ali et al., 2016; Al-Bassam, 2017).

Blockchain's adoption increased alongside challenges like scalability, energy consumption, and regulatory concerns. Responses included alternative consensus mechanisms like proof-of-stake and private blockchains tailored for enterprise needs.

This flexibility fostered efficient ecosystems for various industries. Beyond crypto-currencies, finance, supply chain, healthcare, and real estate explored blockchain's benefits. It pledged swift cross-border transactions, fraud reduction, and transparency in finance, while also offering traceability in supply chains and enhanced data security in healthcare. Governments recognized its potential for public services, from identity management to voting systems, with varying degrees of adoption and exploration (Bustamante et al., 2022; He et al., 2023; Islam, 2023; Kassen, 2022; Tabatabaei et al., 2023).

7.3 ARTIFICIAL INTELLIGENCE APPLICATIONS IN AIRCRAFT MAINTENANCE

Artificial Intelligence (AI) has revolutionized various industries, and aircraft maintenance management is no exception. With the increasing complexity of modern aircraft systems and the critical importance of safety in aviation, AI-driven applications are being harnessed to enhance efficiency, accuracy, and predictive capabilities in aircraft maintenance processes. These applications span from data analysis to decision-making, resulting in reduced downtime, improved operational performance, and enhanced safety in the aviation sector.

One of the primary areas where AI finds significant application in aircraft maintenance management is predictive maintenance (Safoklov et al., 2022). Traditional maintenance practices often rely on fixed schedules or reactive approaches, leading to unnecessary maintenance activities and potential safety risks. AI enables a shift towards predictive maintenance, where data from sensors, flight logs, and various aircraft systems are analyzed to predict when maintenance is actually required. By utilizing machine learning algorithms, AI can identify patterns of component degradation and predict potential failures before they occur. This not only prevents unplanned downtime but also optimizes maintenance scheduling, reducing costs associated with unnecessary part replacements and maximizing aircraft utilization.

Furthermore, AI enhances fault detection and diagnosis processes. Modern aircraft generate huge volumes of data during flights, encompassing parameters such as engine performance, avionics readings, and sensor outputs. AI algorithms can rapidly analyze this data in real-time to identify anomalies or deviations from expected behaviour. This helps maintenance teams to quickly pinpoint the root causes of issues and initiate appropriate corrective actions. By automating the analysis of complex data, AI expedites the decision-making process, leading to reduced turnaround times for maintenance tasks (Mast et al., 1999; Mian et al., 2023; Zhu et al., 2022).

Closely related to fault detection is health monitoring. AI-driven health monitoring systems continuously assess the condition of critical components and systems in real-time. For instance, sensors embedded within engines can transmit data to AI algorithms that monitor performance parameters. If any parameter falls outside normal ranges, the system can trigger alerts, allowing maintenance teams to proactively address potential problems. This approach significantly enhances aviation safety by preventing the operation of aircraft with compromised components (Aghazadeh Ardebili et al., 2023; Naik et al., 2023).

AI also facilitates the management of maintenance records and documentation. Keeping track of maintenance histories for different aircraft, components, and systems can be overwhelming. AI-powered solutions can organize and manage this data efficiently, ensuring that maintenance records are accurate, up-to-date, and easily accessible. This is particularly valuable for regulatory compliance and auditing purposes.

In addition to these technical applications, AI can streamline communication and collaboration within maintenance teams. Virtual assistants powered by AI can provide technicians with instant access to information and procedural guidelines. This reduces the time spent searching for information and helps standardize maintenance procedures, thereby enhancing consistency and reducing the likelihood of errors.

Nonetheless, the integration of AI in aircraft maintenance management is not without challenges. Data quality and security are paramount, as inaccurate or compromised data can lead to erroneous predictions or unsafe conditions. Therefore, establishing robust data collection and validation processes is crucial. Moreover, while AI can assist in decision-making, human expertise remains essential for interpreting complex situations and making final judgements. Training maintenance personnel to work effectively alongside AI systems is essential for maximizing their benefits.

In conclusion, the application of artificial intelligence in aircraft maintenance management is transforming the aviation industry. By enabling predictive maintenance, enhancing fault detection, automating health monitoring, and streamlining documentation, AI improves operational efficiency, reduces costs, and enhances safety. The ability to predict component failures before they occur not only minimizes unscheduled downtime but also optimizes the utilization of aircraft fleets. As technology continues to advance and AI systems become more sophisticated, the aviation sector is poised to experience even greater benefits from AI-driven solutions. However, a balanced approach that combines AI capabilities with human expertise will be pivotal in ensuring the continued safety and reliability of air travel.

7.4   AN OVERVIEW OF MACHINE LEARNING ALGORITHMS

In this section some of the state-of-the-art machine learning algorithms frequently used in predictive modelling have been presented. Machine learning techniques can be grouped into two algorithm categories: regression and classification. Artificial neural networks and gradient boosting algorithms are some of the most effective methods used in practice. These methods are also closely related to optimization and therefore the application of efficient optimization algorithms plays a crucial role in machine learning implementation. Optimization algorithms are mainly used in hyperparameter tuning for gradient boosting algorithms and in weight optimization for artificial neural networks. As a generic example of artificial neural networks Figure 7.1 shows the input nodes, weights, hidden layer and output layer of an artificial neural network. In Figure 7.1, each node is split into two parts denoted with z and a, where z consists of a linear combination of the input values, and a is the output of the activation function. Commonly used activation functions are the sigmoid function and the arctangent function. In a

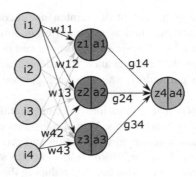

FIGURE 7.1   Example of an artificial neural network.

supervised learning task, the weights w_{ij} and g_{ij} are updated based on the accuracy of the model output using the backpropagation algorithm. As an alternative to the backpropagation method, the problem of minimizing the prediction error of a neural network is often treated as an optimization task. In recent years, many newly developed metaheuristic optimization techniques such as harmony search (Cakiroglu et al., 2023b; Geem et al., 2001), social spider optimization (Cakiroglu et al., 2021a; 2021b), Jaya optimization (Cakiroglu et al., 2022), manta ray foraging (Cakiroglu et al., 2023a; Zhao et al., 2020), arithmetic optimization algorithm (Abualigah et al., 2021), reptile search algorithm (Abualigah et al., 2022) have been utilized in the optimization of various engineering systems. These metaheuristic algorithms have also been implemented in the weight optimization of artificial neural networks with the objective of minimizing the deviation of the network output from the known target values.

$$E = \frac{1}{2}\left(t - a_4\right)^2 \Rightarrow \frac{\partial E}{\partial a_4} = a_4 - t, \quad a_4 = \frac{1}{1 + e^{-z_4}} \Rightarrow \frac{\partial a_4}{\partial z_4} = a_4\left(1 - a_4\right) \quad (7.1)$$

The error E of a neural network is calculated according to Eq. 7.1 where t is the known target value that is being predicted by the network. Eq. 7.1 shows the calculation of the node output a_4 by the sigmoid function. In the backpropagation algorithm the weights are being updated according to the formula in Eq. 7.2 where α is the learning rate which could be any real number in the interval $(0, 1)$. The learning rate is a hyperparameter of the algorithm and the exact value of t which leads to best performance could be obtained from optimization.

$$w_{ij} := w_{ij} - \alpha \frac{\partial E}{\partial w_{ij}} \quad (7.2)$$

The partial derivative term in Eq. 7.2 is obtained by the chain rule. An example of this procedure is demonstrated in Eq. 7.3 and Eq. 7.4 for the update of the weight w_{11}.

$$w_{11} := w_{11} - \alpha \frac{\partial E}{\partial w_{11}} \quad (7.3)$$

$$\frac{\partial E}{\partial w_{11}} = \frac{\partial E}{\partial a_4} \frac{\partial a_4}{\partial z_4} \frac{\partial z_4}{\partial a_1} \frac{\partial a_1}{\partial z_1} \frac{\partial z_1}{\partial w_{11}} \tag{7.4}$$

Inserting the expressions for the partial derivatives of E, a_4 and a_1 from Eq. 7.1 into Eq. 7.4, Eq. 7.4 can be rewritten as in Eq. 7.5.

$$\frac{\partial E}{\partial w_{11}} = (a_4 - t) a_4 (1 - a_4) \frac{\partial z_4}{\partial a_1} a_1 (1 - a_1) \frac{\partial z_1}{\partial w_{11}} \tag{7.5}$$

If we introduce the variable δ_1 for $\dfrac{\partial E}{\partial z_1}$, we can rewrite Eq. 5.5 as in Eq. 7.6.

$$\frac{\partial E}{\partial w_{11}} = \delta_1 \frac{\partial z_1}{\partial w_{11}} \tag{7.6}$$

According to Figure 7.1, z_1 is equal to $i_1 w_{11} + i_2 w_{21} + i_3 w_{31} + i_4 w_{41}$ and $z_4 = a_1 g_{14} + a_2 g_{24} + a_3 g_{34}$. Inserting these expressions for z_1 and z_4 in Eq. 5.5 and Eq. 7.6 we obtain Eq. 7.7.

$$\frac{\partial E}{\partial w_{11}} = \delta_1 i_1, \quad \delta_1 = \frac{\partial E}{\partial z_1} = (a_4 - t) a_4 (1 - a_4) g_{14} a_1 (1 - a_1) \tag{7.7}$$

In a similar way the partial derivatives of the error function with respect to the weights of the network can be obtained as in Eq. 7.8 to Eq. 7.12.

$$\frac{\partial E}{\partial w_{12}} = \delta_2 i_1, \quad \delta_2 = \frac{\partial E}{\partial z_2} = (a_4 - t) a_4 (1 - a_4) g_{24} a_2 (1 - a_2) \tag{7.8}$$

$$\frac{\partial E}{\partial w_{13}} = \delta_3 i_1, \quad \delta_3 = \frac{\partial E}{\partial z_3} = (a_4 - t) a_4 (1 - a_4) g_{34} a_3 (1 - a_3) \tag{7.9}$$

$$\frac{\partial E}{\partial w_{21}} = \delta_1 i_2, \quad \frac{\partial E}{\partial w_{22}} = \delta_2 i_2, \quad \frac{\partial E}{\partial w_{23}} = \delta_3 i_2 \tag{7.10}$$

$$\frac{\partial E}{\partial w_{31}} = \delta_1 i_3, \quad \frac{\partial E}{\partial w_{32}} = \delta_2 i_3, \quad \frac{\partial E}{\partial w_{33}} = \delta_3 i_3 \tag{7.11}$$

$$\frac{\partial E}{\partial w_{41}} = \delta_1 i_4, \quad \frac{\partial E}{\partial w_{42}} = \delta_2 i_4, \quad \frac{\partial E}{\partial w_{43}} = \delta_3 i_4 \tag{7.12}$$

Another frequently used method of machine learning is the gradient boosting or the ensemble learning methodology. This technique combines the predictions of multiple decision trees in order to obtain a more accurate prediction. Among these methods extreme gradient boosting (XGBoost), light gradient boosting machine (LightGBM), categorical boosting (CatBoost), and random forest. This procedure can be visualized as in Figure 7.2 where each decision tree corrects the errors done by the previous decision trees and the final model output is the sum of all decision tree predictions (Aladsani et al., 2022).

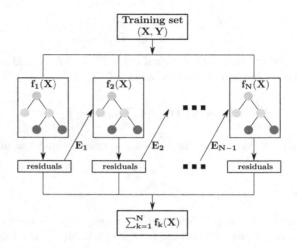

FIGURE 7.2 Ensemble learning process.

In Figure 7.2 the individual decision tree predictors are denoted with $f_i(X)$ where X is the part of the dataset used for training the predictor trees and $i \in [1, N]$ with N denoting the total number of decision trees in the ensemble learning model. The errors in the predictions of the individual decision trees are shown with E_i in Figure 7.2. The ensemble learning prediction can be summarized as in Eq. 7.13, where $\mathbf{x}$ is an arbitrary data point and $\hat{y}$ is the predicted value for the target y (Cakiroglu et al., 2023c).

$$f(\mathbf{x}) = \sum_{k=1}^{N} f_k(\mathbf{x}) = \hat{y} \tag{7.13}$$

Interpretability of machine learning algorithms can be achieved by the implementation of the SHAP methodology. The SHAP algorithm supplies a visual representation of the feature impacts on the output of a predictive model. SHAP is an acronym for SHapley Additive exPlanations, a technique used in machine learning to determine how input variables contribute to output predictions. It is based on game theory and the concept of Shapley values, which is a formula for understanding how much each participant contributes to a cooperative game. In the context of machine learning, the players in the cooperative game are the input features, and the objective is to solve the prediction problem. SHAP estimates the impact of each feature by analyzing variations in model outputs when a feature is included or excluded from the input. SHAP defines an explanation model as a linear combination of input features, as demonstrated in Eq. 7.14.

$$s(x') = \phi_0 + \sum_{k=1}^{N} \phi_k x'_k \tag{7.14}$$

In Eq. 7.14 ϕ_k are the Shapley values, and N is the total number of data points in the dataset. In Eq. 7.14, s denotes an explanation model and x' is a simplified input vector. A mapping function f relates the actual data points in the dataset to the simplified vectors such that $x = f(x')$. The Shapley values ϕ_i can be computed as in Eq. 7.15 (Mangalathu et al., 2020).

$$\phi_i = \sum_{S \subseteq F \backslash \{i\}} \frac{|S|!(|F| - |S| - 1)!}{|F|!} \left[g_{S \cup \{i\}} \left(x_{S \cup \{i\}} \right) - g_S \left(x_S \right) \right]$$ (7.15)

In Eq. 7.15, F is the set of all input features and S is a subset of F where the feature with index i is not included.

7.5   A CASE STUDY OF THE APPLICATION OF BLOCKCHAIN TECHNOLOGY IN AIRCRAFT MAINTENANCE MANAGEMENT

The study of Efthymiou et al. (2022) aimed to explore the potential of blockchain technology in maintenance, repair, and overhaul (MRO) organizations. The researchers used semi-structured interviews and a single case study to understand the process of maintenance verification and records management in MRO. The study found that blockchain technology could bring several benefits to the aviation industry, including increased transparency, trust, and security in exchanging information. The study also identified some barriers to implementing blockchain technology in MRO, such as the lack of standardization and regulatory frameworks.

Dublin Aerospace is a leading MRO organization in Ireland that provides maintenance services for aircraft and components such as the overhaul of auxiliary power units, landing gears, and integrated drive generators. Currently, the company holds approvals from United States Federal Aviation Administration (FAA), The European Authority for Aviation Safety (EASA), Transport Canada Civil Aviation (TCCA), Civil Aviation Authority of the United Kingdom (CAA), Egyptian Civil Aviation Authority (ECAA), and the Civil Aviation Authorities of Thailand, Cayman Islands, Guernsey, and Bermuda. As of 2023, the company has the capacity to carry out base maintenance for about 70 A320, A330, and 737 type aircraft and to overhaul about 200 auxiliary power units per year. Based on extensive interviews with Dublin Aerospace and facility inspections, the existing maintenance verification and records management methods have been compiled. As part of the maintenance procedure for the aircraft and components they service, Dublin Aerospace keeps track of the repair tasks performed. In accordance with their contractual duties, they supply these records in both hard and soft versions. The company has a paper record called the original document that they feed into their maintenance system. Every repair and inspection performed on the aircraft, auxiliary power unit, and landing gears is documented by their maintenance system. Efthymiou et al. (2022) conducted semi-structured interviews and a survey with key stakeholders from Dublin Aerospace to explore their views on Blockchain technology and its potential benefits and challenges for MRO. The case study revealed that Dublin Aerospace has a high level of awareness and interest in Blockchain technology, but also faces some barriers to adoption, such as lack of standardization, regulation, trust, and collaboration amongst industry partners.

By keeping track of every installation and removal of an aviation part, blockchain technology could be applied to MRO services. This can increase system efficiency and confidence by facilitating stakeholders' easy access to aircraft maintenance records. Additionally, the use of blockchain technology can improve information exchange security, transparency, and trust. Stakeholders can obtain a real-time picture of an airplane's state with blockchain technology, including information on a part's provenance and history as well as the credentials and identity of the personnel who installed or repaired it.

7.6 CONCLUSIONS

By offering an unchangeable digital record of all planned and unforeseen maintenance actions performed on every aircraft, blockchain technology has the potential to completely transform the way that airplanes are maintained. It can increase worker productivity while lowering expenses associated with unscheduled maintenance and downtime. In order to promote confidence and openness in the information exchange process, the blockchain might be utilized to securely store records of aircraft maintenance.

In this study, the potential use cases of the blockchain technology in aircraft maintenance have been mentioned. Following a background introduction on the applications of machine learning methodologies in aviation, an overview of the state-of-the-art machine learning methodologies such as artificial neural networks and ensemble learning methods has been presented. Fault detection and diagnosis in aircraft maintenance is an area where machine learning techniques can improve efficiency. By using machine learning algorithms to analyze data from sensors on an aircraft, early fault detection can be achieved. Using blockchain technology, digital twins, predictive maintenance, fault detection and diagnosis, efficiency in aircraft maintenance can be significantly improved. By providing real-time information about an aircraft's condition, these technologies can help reduce downtime, improve safety, and increase productivity.

REFERENCES

Abualigah, L., Diabat, A., Mirjalili, S., Abd Elaziz, M., & Gandomi, A. H. (2021). The arithmetic optimization algorithm. *Computer Methods in Applied Mechanics and Engineering*, 376, 113609.

Abualigah, L., Abd Elaziz, M., Sumari, P., Geem, Z. W., & Gandomi, A. H. (2022). Reptile Search Algorithm (RSA): A nature-inspired meta-heuristic optimizer. *Expert Systems with Applications*, 191, 116158.

Aghazadeh Ardebili, A., Ficarella, A., Longo, A., Khalil, A., & Khalil, S. (2023). Hybrid turbo-shaft engine digital twinning for autonomous aircraft via AI and synthetic data generation. *Aerospace*, 10(8), 683, https://doi.org/10.3390/aerospace10080683

Ahmed, W. A., & MacCarthy, B. L. (2023). Blockchain-enabled supply chain traceability– How wide? How deep?. *International Journal of Production Economics*, 263, 108963, https://doi.org/10.1016/j.ijpe.2023.108963

Akhmatova, M. S., Deniskina, A., Akhmatova, D. M., & Kapustkina, A. (2022). Green SCM and TQM for reducing environmental impacts and enhancing performance in the aviation spares supply chain. *Transportation Research Procedia*, 63, 1505–1511, https://doi.org/10.1016/j.trpro.2022.06.162

Aladsani, M. A., Burton, H., Abdullah, S. A., & Wallace, J. W. (2022). Explainable machine learning model for predicting drift capacity of reinforced concrete walls. *ACI Structural Journal*, 119(3), https://doi.org/10.14359/51734484

Al-Bassam, M. (2017, April). SCPKI: A smart contract-based PKI and identity system. In *Proceedings of the ACM Workshop on Blockchain, Cryptocurrencies and Contracts* (pp. 35–40).

Alharby, M. & van Moorsel, A. (2017). Blockchain based smart contracts: A systematic mapping study. In *2018 International Conference on Cloud Computing, Big Data And Blockchain (ICCBB)* (pp. 125–140), https://doi.org/10.5121/csit.2017.71011

Ali, M., Nelson, J., Shea, R., & Freedman, M. J. (2016). Blockstack: A global naming and storage system secured by blockchains. In *2016 USENIX Annual Technical Conference (USENIX ATC 16)* (pp. 181–194).

Ashraf, M., & Heavey, C. (2023). A prototype of supply chain traceability using Solana as blockchain and IoT. *Procedia Computer Science*, 217, 948–959.

Bustamante, P., Cai, M., Gomez, M., Harris, C., Krishnamurthy, P., Law, W.,... Weiss, M. (2022). Government by code? Blockchain applications to public sector governance. *Frontiers in Blockchain*, 5, 869665, https://doi.org/10.3389/fbloc.2022.869665

Cakiroglu, C., Islam, K., Bekdaş, G., & Billah, M. (2021a). CO_2 emission and cost optimization of concrete-filled steel tubular (CFST) columns using metaheuristic algorithms. *Sustainability*, 13, 8092, https://doi.org/10.3390/su13148092

Cakiroglu, C., Islam, K., Bekdaş, G., Kim, S., & Geem, Z. W. (2021b). CO_2 emission optimization of concrete-filled steel tubular rectangular stub columns using metaheuristic algorithms. *Sustainability*, 13, 10981, https://doi.org/10.3390/su131910981.

Cakiroglu, C., Islam, K., Bekdaş, G., & Apak, S. (2022). Cost and CO_2 emission-based optimisation of reinforced concrete deep beams using Jaya algorithm. *Journal of Environmental Protection and Ecology*, 23(6), 2420–2429.

Cakiroglu, C., Islam, K., & Bekdaş, G. (2023a). Manta Ray foraging and Jaya hybrid optimization of concrete filled steel tubular stub columns based on CO_2 emission. In Bekdaş, G., & Nigdeli, S.M. (Eds.), *Hybrid Metaheuristics in Structural Engineering. Studies in Systems, Decision and Control*, vol. 480. Springer, Cham, https://doi.org/10.1007/978-3-031-34728-3_7

Cakiroglu, C., Islam, K., Bekdaş, G., & Nehdi, M. L. (2023b). Data-driven ensemble learning approach for optimal design of cantilever soldier pile retaining walls. In *Structures*, vol. 51 (pp. 1268–1280). Elsevier, https://doi.org/10.1016/j.istruc.2023.03.109

Cakiroglu, C., Shahjalal, M., Islam, K., Mahmood, S. F., Billah, A. M., & Nehdi, M. L. (2023c). Explainable ensemble learning data-driven modeling of mechanical properties of fiber-reinforced rubberized recycled aggregate concrete. *Journal of Building Engineering*, 76, 107279, https://doi.org/10.1016/j.jobe.2023.107279.

Efthymiou, M., McCarthy, K., Markou, C., & O'Connell, J. F. 2022. An exploratory research on blockchain in aviation: The case of maintenance, repair and overhaul (MRO) organizations. *Sustainability*, 14, 2643, https://doi.org/10.3390/su14052643

Far, S. B., Rad, A. I., & Asaar, M. R. (2023). Blockchain and its derived technologies shape the future generation of digital businesses: A focus on decentralized finance and the Metaverse. *Data Science and Management*, 6(3), 183–197, https://doi.org/10.1016/j.dsm.2023.06.002

Geem, Z. W., Kim, J. H., & Loganathan, G. V. (2001). A new heuristic optimization algorithm: Harmony search. *Simulation*, 76(2), 60–68, https://doi.org/10.1177/003754970107600201.

He, C., Tan, C., Ip, W. H., & Wu, C. H. (2023). Combating counterfeits with the blockchain-technology-supported platform under government enforcement. *Transportation Research Part E: Logistics and Transportation Review*, 175, 103155, https://doi.org/10.1016/j.tre.2023.103155

Hewa, T., Ylianttila, M., & Liyanage, M. (2021). Survey on blockchain based smart contracts: Applications, opportunities and challenges. *Journal of Network and Computer Applications*, 177, 102857, https://doi.org/10.1016/j.jnca.2020.102857

Ho, G. T., Tang, Y. M., Tsang, K. Y., Tang, V., & Chau, K. Y. (2021). A blockchain-based system to enhance aircraft parts traceability and trackability for inventory management. *Expert Systems with Applications*, 179, 115101, https://doi.org/10.1016/j.eswa.2021.115101

Huang, S., Wang, G., Yan, Y., & Fang, X. (2020). Blockchain-based data management for digital twin of product. *Journal of Manufacturing Systems*, 54, 361–371, https://doi.org/10.1016/j.jmsy.2020.01.009.

Islam, M. D. (2023). A survey on the use of blockchains to achieve supply chain security. *Information Systems*, 102232, https://doi.org/10.1016/j.is.2023.102232.

Kassen, M. (2022). Blockchain and e-government innovation: Automation of public information processes. *Information Systems*, 103, 101862, https://doi.org/10.1016/j.is.2021.101862

Kuhle, P., Arroyo, D., & Schuster, E. (2021). Building A blockchain-based decentralized digital asset management system for commercial aircraft leasing. *Computers in Industry*, 126, 103393, https://doi.org/10.1016/j.compind.2020.103393

Li, Y., & Mardani, A. (2023). Digital twins and blockchain technology in the industrial Internet of Things (IIoT) using an extended decision support system model: Industry 4.0 barriers perspective. *Technological Forecasting and Social Change*, 195, 122794, https://doi.org/10.1016/j.techfore.2023.122794

Lopes, D. P., Rita, P., & Treiblmaier, H. (2021). The impact of blockchain on the aviation industry: Findings from a qualitative study. *Research in Transportation Business & Management*, 41, 100669, https://doi.org/10.1016/j.rtbm.2021.100669

Mandolla, C., Petruzzelli, A. M., Percoco, G., & Urbinati, A. (2019). Building a digital twin for additive manufacturing through the exploitation of blockchain: A case analysis of the aircraft industry. *Computers in industry*, 109, 134–152, https://doi.org/10.1016/j.compind.2019.04.011

Mangalathu, S., Hwang, S. H., & Jeon, J. S. (2020). Failure mode and effects analysis of RC members based on machine-learning-based SHapley Additive exPlanations (SHAP) approach. *Engineering Structures*, 219, 110927, https://doi.org/10.1016/j.engstruct.2020.110927

Mast, T. A., Reed, A. T., Yurkovich, S., Ashby, M., & Adibhatla, S. (1999, August). Bayesian belief networks for fault identification in aircraft gas turbine engines. In *Proceedings of the 1999 IEEE International Conference on Control Applications (Cat. No. 99CH36328)*, vol. 1 (pp. 39–44). IEEE, https://doi.org/10.1109/CCA.1999.806140

Mian, T., Choudhary, A., Fatima, S., & Panigrahi, B. K. (2023). Artificial intelligence of things based approach for anomaly detection in rotating machines. *Computers and Electrical Engineering*, 109, 108760, https://doi.org/10.1016/j.compeleceng.2023.108760

Naik, J., Acharya, A., & Thaker, J. (2023). Revolutionizing condition monitoring techniques with integration of artificial intelligence and machine learning. *Materials Today: Proceedings*, https://doi.org/10.1016/j.matpr.2023.08.262

Rolinck, M., Gellrich, S., Bode, C., Mennenga, M., Cerdas, F., Friedrichs, J., & Herrmann, C. (2021). A concept for blockchain-based LCA and its application in the context of aircraft MRO. *Procedia CIRP*, 98, 394–399, https://doi.org/10.1016/j.procir.2021.01.123

Safoklov, B., Prokopenko, D., Deniskin, Y., & Kostyshak, M. (2022). Model of aircraft maintenance repair and overhaul using artificial neural networks. *Transportation Research Procedia*, 63, 1534–1543, https://doi.org/10.1016/j.trpro.2022.06.165

Santonino, M. D., Koursaris, C. M., & Williams, M. J. (2018). Modernizing the supply chain of Airbus by integrating RFID and blockchain processes. *International Journal of Aviation, Aeronautics, and Aerospace*, 5(4), 4, https://doi.org/10.15394/ijaaa.2018.1265

Sasikumar, A., Vairavasundaram, S., Kotecha, K., Indragandhi, V., Ravi, L., Selvachandran, G., & Abraham, A. (2023). Blockchain-based trust mechanism for digital twin empowered Industrial Internet of Things. *Future Generation Computer Systems*, 141, 16–27, https://doi.org/10.1016/j.future.2022.11.002

Schultz, M., Evler, J., Asadi, E., Preis, H., Fricke, H., & Wu, C. L. (2020). Future aircraft turnaround operations considering post-pandemic requirements. *Journal of Air Transport Management*, 89, 101886, https://doi.org/10.1016/j.jairtraman.2020.101886.

Tabatabaei, M. H., Vitenberg, R., & Veeraragavan, N. R. (2023). Understanding blockchain: Definitions, architecture, design, and system comparison. *Computer Science Review*, 50, 100575, https://doi.org/10.1016/j.cosrev.2023.100575

Zhao, W., Zhang, Z., & Wang, L. (2020). Manta ray foraging optimization: An effective bio-inspired optimizer for engineering applications. *Engineering Applications of Artificial Intelligence*, 87, 103300.

Zhu, Z., Lei, Y., Qi, G., Chai, Y., Mazur, N., An, Y., & Huang, X. (2022). A review of the application of deep learning in intelligent fault diagnosis of rotating machinery. *Measurement*, 112346, https://doi.org/10.1016/j.measurement.2022.112346

8 Augmented Reality and Virtual Reality Solutions in Aviation Industry

Basar Koc
Stetson University, DeLand, FL, USA

8.1 INTRODUCTION

Extended reality (XR) technologies are widely used in applications to enhance user experience and support other technologies (Cárdenas-Robledo et al., 2022). XR consists of virtual reality (VR), augmented reality (AR), and mixed reality technologies. The VR systems create an artificial environment for their users and usually provide very limited possibilities for the user to interact with surrounding real objects in the real-world environment. While VR is generally the preferred technology in the gaming industry and has many advantages, such as VR devices being generally more affordable for end-users, VR technology has certain disadvantages and limitations. Suppose the user needs to interact and use the surrounding objects in the real world with low-powered headsets currently available. In that case, relaying the real-world environment to a virtual one is impossible. To overcome this problem, augmented reality (AR) technology was introduced in the 1990s, and it was used for the first time in a commercial application by Boeing in 2015. The AR technology integrates both real and virtual environments and allows users to see and interact with objects in both real and virtual environments. AR technology can be used on any device with a camera and display. The third group of technology is mixed reality. In mixed reality, both environments merge into a single (mixed) environment, as shown in Figure 8.1. To experience mixed reality, wearing a headset is required. Oculus Quest, Oculus Rift, Google Glass, Sony PlayStation VR, Samsung Gear, HTC Vive, Microsoft HoloLens, Steam Valve VR Kit, and Apple Vision Pro are some of the currently available extended reality headsets on the market.

These technologies present many business solutions that lead to operational excellence for the aviation sector. Both technologies help aviation companies to provide better service and train their staff efficiently. Some AR and VR applications in the aviation industry include aircraft assembly, maintenance, ramp handling operations, communication, data collection, personnel training, indoor navigation for airports, and passenger entertainment. This chapter discusses AR and VR solutions and their applications in the aviation industry by introducing real-life smart applications. The remainder of this chapter is organized as follows: Section 8.2 discusses the use cases of augmented and virtual reality technologies. In Section 8.3, we provide an overview of the AR and VR applications developed by the companies in the industry.

106

DOI: 10.1201/9781003389187-8

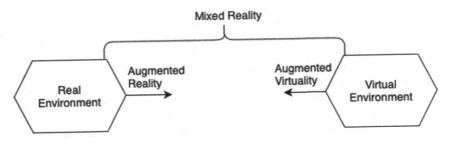

FIGURE 8.1 The reality-virtuality (RV) continuum.

(Milgram et al., 1995.)

Finally, in Section 8.4, a summary of these technologies in the aviation industry is given.

8.2 THE USE CASES OF AUGMENTED AND VIRTUAL REALITY IN AVIATION INDUSTRY

Many companies actively use XR technologies in their daily operations. Demir et al. (2020) discussed the use cases of AR devices in supply chain management and how AR integration into smart factories helped to improve supply chain management at large warehouses, as shown in Table 8.1. The aircraft manufacturing companies, such as Embraer, Airbus, and Boeing, utilize these technologies in designing and maintaining the aircraft, as summarized by Frigo et al. (2016). In this section, we will visit the use cases of VR and AR in the aviation industry as follows:

- Education: VR and AR technologies can be used to train pilots, cabin crew, and maintenance engineers. Flight simulations can be performed on VR headsets for unique emergency scenarios in a safe, controlled environment. Virtual reality can improve the quality of the training for maintenance

TABLE 8.1
An Overview of Augmented Reality Use Cases in Manufacturing

Phase	Company	Rendering Technology	Improvement Quantifications
Production Operations	Lockheed Martin	Smart glasses	Increases engineers' accuracy to 96%, increases work speed by 30%
Production Operations	Airbus	Google glasses	Cut production times by one-fourth, cut error rates by half
Production Operations	Boeing	Microsoft HoloLens	Not quantified
Inspections & Error Prevention	Airbus	Tablet	Time reduction of 80%

Adapted from Kohn and Harborth (2018).

personnel (Gómez-Cambronero, 2023). It can also be employed in training air traffic controllers (ATC), which requires practicing a wide variety of realistic traffic scenarios and finding solutions in various chaotic situations. For instance, Bowling et al. (2008) showed that VR environments can be used in the training of aircraft maintenance inspectors. Meister et al. (2022) proposed an augmented reality system for enhancing general aviation weather education. The research focused on thunderstorm cell lifecycle visualization to improve the in-flight decision-making process. Lee et al. (2022) introduced a system for integrating VR technology into classrooms and tested the proposed technique with an aircraft maintenance simulation. Lufthansa's training programme and other airlines, such as KLM, now involve VR training courses for its cabin crew.

- Manufacturing: VR and AR headsets are used in aircraft design and manufacturing processes, for instance, Embraer (2019) and Microsoft (2020). Various interior configurations can be applied virtually using a headset, and the technology helps the designers test different layout configurations. A well-designed cockpit increases the safety of the flight.
- Assistance: These technologies can provide certain real-time information, such as weather, navigation, and terrain information, for pilots to enhance flight safety. This data helps pilots in the decision-making process during a flight. ATC can utilize the technology to overlay flight paths and aircraft locations on their screens. In an emergency during a flight, a maintenance engineer or specialized technician provides support to the pilot remotely via a virtual headset to diagnose the problem from the ground and provide the required assistance. For instance, Gorbunov et al. (2015) presented a repurposed head-up display (HUD) for pilots. The stereoscopic version of the tunnel-in-the-sky system was applied to augmented reality devices.
- Maintenance: The aircraft manufacturers determine the aircraft's maintenance schedule, and aviation safety agencies strictly enforce this schedule. Performing repair and maintenance on the aircraft requires following a procedure and instructions published by the manufacturer. AR technology helps engineers and technicians follow step-by-step instructions and perform the necessary repairs. The use of technology decreases the repair time and downtime of aircraft. Boeing and Airbus developed an AR app for their technicians to provide maintenance procedures for their aircraft. Vora et al. (2002) presented that using VR technologies in maintenance is effective using a system developed for a visual inspection task of an aft cargo bay. Eschen et al. (2018) developed a framework for mixed reality systems to evaluate the inspection and maintenance process in the aviation industry. Peng et al. (2022) designed an augmented reality application on the Microsoft HoloLens 2 platform for aviation equipment inspection and maintenance training. In the proposed application, the artificial system was also employed to extend the effectiveness of the training process further and reduce the time that is needed to diagnose and fix the problems on the aircraft. Wu and Van-Hoan (2022) presented a virtual reality system to inspect

the Dornier-228 aircraft's fuel system visually. The students in the training programme validated the proposed system's effectiveness and found that the VR system is highly effective.

- Improving passenger experience: The VR and AR technologies can provide real-time information such as flight updates, gates, and entertainment options for passengers. One of the concerns of using these technologies is getting passenger motion sickness while wearing headsets in flight. In Soyka et al. 2015, studied the effects of turbulent motions on motion sickness. The participants' subjective opinions and physiological signals, such as their vitals, were evaluated in the experiments to measure the effects of three different turbulent scenarios. The researchers concluded that no severe motion sickness was connected to passengers wearing virtual reality headsets in flight. The other concern is wearing these headsets in public. The social acceptability of wearing VR headsets in-flight and how usable the VR technology as an in-flight entertainment device for passengers was studied by Williamson et al. (2019). Similarly, Ng et al. (2021) studied the passenger experience of mixed-reality virtual display layouts in airplane environments. The research findings suggested that passengers prefer virtual layouts over airplanes' physical displays.

8.3 CASE STUDY: THE AR AND VR APPLICATIONS IN AVIATION INDUSTRY

The use of AR and VR devices has been actively investigated to use in the aviation industry for manufacturing, training, and entertainment purposes. In this section, we briefly summarize the efforts of Embraer, KLM, and NASA in using AR and VR technologies in aviation.

8.3.1 VR APPLICATIONS IN EMBRAER

The Brazilian aerospace company Embraer established its first Virtual Reality Center (Centro de Realidade Virtual—CRV) in February 2000 in São José dos Campos, São Paulo. The centre includes a wide variety of tools that enable the company to keep the highest quality in manufacturing by eliminating flaws in assembly. VR technology helped the company prevent problems in the design and production of the aircraft by letting the engineers solve the issues before assembling the parts, as shown in Figure 8.2. In an article published by Embraer in 2019, the Engineering and Manufacturing Digital Solutions specialist Fábio de Matos Leite states that "the CRV can be used in all stages of aircraft development, but it's more used in the stage that precedes the first flight" (Embraer, 2019). The technology helped the company reduce the time needed to manufacture its first-generation E-Jet commercial aircraft, Embraer 170. The company developed E170 aircraft using VR technologies in 38 months compared to their previous model, ERJ145, which was completed in 60 months. In 2017, Embraer upgraded its technology centre to take advantage of new studies in the aircraft manufacturing field and continued to use VR technology to develop new E-Jets, E2.

FIGURE 8.2   Embraer Virtual Reality Center and E190 aircraft.

(Image: Embraer.)

8.3.2   VR APPLICATIONS IN KLM

Pilots must take a Type Rating Course and get certified for each aircraft they want to operate. The KLM, Koninklijke Luchtvaart Maatschappij, Dutch Royal Airline Group started using VR devices to train its pilots for Standard Operating Procedures (SOP) during the Type Rating courses in its Cityhopper subsidiary. The KLM Cityhopper group mainly uses Embraer 175 and 190 fleet for its European flights. The pilot training programme uses VR devices developed by the developers at KLM to increase the effectiveness and reduce the cost of the pilot training programme while maintaining high safety standards (KLM, 2020a). Practicing with a virtual cockpit helped the pilots familiarize themselves with the aircraft cockpit as much as they wanted before attending the on-site simulation programme. The VR training programme consisted of a virtual cockpit, an instructional point-of-view (POV) flight video, and a virtual walkaround.

- Virtual cockpit: The pilots could use VR headsets to interact with the virtual cockpit through computer-generated static images of the controller panels inside the Embraer aircraft (KLM, 2020b). The interactive virtual cockpit was designed by VR developers and 3D designers of the KLM, as shown in Figure 8.3.
- Flight video: The pilots had access to a 360-degree jumpseat POV flight video from the aircraft's cockpit without having to operate an actual flight. The video was recorded by a VR engineer of the KLM while sitting in the cockpit near pilots in a flight.
- Virtual walkaround: The pilots were able to check the outside of the aircraft using pre-recorded photographs.

The training programme introduced in 2021 was the first attempt at an airline-integrated VR technology for a fleet of Embraer aircraft. After the success of this training programme, KLM extended the use of VR and AR technologies in training maintenance engineers and cabin crews. The airline continues to consider extending these training programmes and discusses using them in the courses to obtain European Aviation Safety Agency (EASA) certifications.

FIGURE 8.3 KLM Training Programme.

(Image: KLM Corporate Communications.)

8.3.3 AR APPLICATIONS IN NASA

NASA is another organization that started using mixed reality technology in regular maintenance operations on the International Space Station. The project was "Sidekick: Investigate immersive visualization capabilities" and was jointly developed by Microsoft and NASA. On December 6, 2015, the mixed reality headset, HoloLens, was sent to the International Space Station (ISS). In the Sidekick project, the crew members wore mixed-reality Microsoft HoloLens headsets and used them to complete their assigned tasks (Norris, 2015). Each task was pre-programmed and could be viewed holographically by the crew member, as in Figure 8.4. The headset had two modes: remote expert and procedure modes. In the remote expert mode, the headset can provide real-time feedback to crew members from the engineers on the ground using its holographic voice, video, and virtual annotation functionality. With this mode, the experts on the ground could see what a crew member saw and assist

FIGURE 8.4 The concept of the Sidekick programme.

(Image: NASA.)

the crew member in a complex repair scenario. The second mode was procedure mode, designed mainly for offline guidance and instructional purposes. With this mode, the crew members were able to follow pre-programmed instructions on the headset instead of printed complex manuals.

8.3.4 AR APPLICATIONS IN LOCKHEED MARTIN

After the success of the Sidekick project in 2017, American aerospace corporation Lockheed Martin decided to use mixed reality technology in manufacturing components of NASA's Orion spacecraft. The Orion capsule was part of NASA's Artemis programme to carry humans to space (Microsoft, 2020). The Microsoft HoloLens 2 headset was mainly used to install some of the modules inside the spacecraft, requiring high precision to prevent catastrophic failures in flight, as shown in Figure 8.5.

8.4 CONCLUSION

VR and AR technologies are used in many industries to improve the user experience. These technologies are effective tools in education, especially in incremental learning. In the aviation industry, many areas require a wide variety of training. Therefore, aircraft manufacturing companies, such as Boeing, Airbus, and Embraer, use these technologies in designing and manufacturing airplanes. Airline companies, such as KLM and Lufthansa, design new training programmes around VR and AR technologies to provide a better experience for their pilots, cabin crews, ground workers, and maintenance engineers. With the advancements in wearable headset devices, there will be more opportunities to employ these technologies in the aviation industry.

FIGURE 8.5 A Lockheed Martin technician at the NASA assembly facility assembles Orion's crew seats.

(Image: Lockheed Martin.)

REFERENCES

Bowling, Shannon R., Khasawneh, Mohammad T., Kaewkuekool, Sittichai, Jiang, Xiaochun, and Gramopadhye, Anand K.. "Evaluating the effects of virtual training in an aircraft maintenance task." *The International Journal of Aviation Psychology* 18, no. 1 (2008): 104–116.

Cárdenas-Robledo, Leonor Adriana, Hernández-Uribe, Óscar, Reta, Carolina, and Cantoral-Ceballos, José Antonio. "Extended reality applications in industry 4.0. – A systematic literature review." *Telematics and Informatics* 73 (2022): 101863.

Demir, Sercan, Yilmaz, Ibrahim, and Paksoy, Turan. "Augmented reality in supply chain management." In *Logistics 4.0*, pp. 136–145. CRC Press (2020).

Embraer. "How virtual reality speeds up aircraft development." Portal Embraer (2019). https://journalofwonder.embraer.com/global/en/82-how-virtual-reality-speeds-up-aircraft-development (Accessed: August 31, 2023).

Eschen, Henrik, Kötter, Tobias, Rodeck, Rebecca, Harnisch, Martin, and Schüppstuhl, Thorsten. "Augmented and virtual reality for inspection and maintenance processes in the aviation industry." *Procedia Manufacturing* 19 (2018): 156–163.

Frigo, Mauricio A., da Silva, E. C. and Barbosa, G.F. "Augmented reality in aerospace manufacturing: A review." *Journal of Industrial and Intelligent Information* 4, no. 2 (2016): 125–130.

Gómez-Cambronero, Águeda, Miralles, Ignacio, Tonda, Anna, and Remolar, Inmaculada. "Immersive virtual-reality system for aircraft maintenance education: A case study." *Applied Sciences* 13, no. 8 (2023): 5043.

Gorbunov, Andrey L., Terenzi, A., and Graziano Terenzi. "Pocket-size augmented reality system for flight control." In *2015 IEEE Virtual Reality (VR)*, pp. 369–369. IEEE (2015).

KLM. "KLM Cityhopper introduces virtual reality training for pilots." KLM Newsroom (2020a). https://news.klm.com/klm-cityhopper-introduces-virtual-reality-training-for-pilots (Accessed: August 31, 2023).

KLM. "VR training for KLC E175 and E190 pilots." KLM Corporate Communications (2020b). https://www.youtube.com/watch?v=czaEQD2vhBI (Accessed: August 31, 2023).

Kohn, Vanessa, and Harborth, David. "Augmented reality – A game changing technology for manufacturing processes?" In *26th European Conference on Information Systems: Beyond Digitization-Facets of Socio-Technical Change (ECIS)*, p. 111. (2018).

Langston, Jennifer. "To the moon and beyond: How HoloLens 2 is helping build NASA's Orion spacecraft." Microsoft Innovation News (2020). https://news.microsoft.com/source/features/innovation/hololens-2-nasa-orion-artemis (Accessed August 31, 2023).

Lee, Hyeonju, Woo, Donghyun, and Yu, Sunjin. "Virtual reality metaverse system supplementing remote education methods: Based on aircraft maintenance simulation." *Applied Sciences* 12, no. 5 (2022): 2667.

Meister, Philippe, Miller, Jack, Wang, Kexin, Dorneich, Michael C., Winer, Eliot, Brown, Lori J., and Whitehurst, Geoffrey. "Designing three-dimensional augmented reality weather visualizations to enhance general aviation weather education." *IEEE Transactions on Professional Communication* 65, no. 2 (2022): 321–336.

Milgram, Paul, Takemura, Haruo, Utsumi, Akira, and Kishino, Fumio. "Augmented reality: A class of displays on the reality-virtuality continuum." In *Telemanipulator and Telepresence Technologies*, vol. 2351, pp. 282–292. SPIE, (1995).

Ng, Alexander, Medeiros, Daniel, McGill, Mark, Williamson, Julie, and Brewster, Stephen. "The passenger experience of mixed reality virtual display layouts in airplane environments." In *2021 IEEE International Symposium on Mixed and Augmented Reality (ISMAR)*, pp. 265–274. IEEE (2021).

Norris, Jeffrey, "Sidekick: Investigating immersive visualization capabilities", Space Station Research Explorer on NASA.gov (2015). https://www.nasa.gov/mission_pages/station/research/experiments/explorer/Investigation.html?#id=2018 (Accessed: August 31, 2023).

Peng, Chao-Chung, Chang, Ai-Chi, and Chu, Yu-Lun. "Application of augmented reality for aviation equipment inspection and maintenance training." In *2022 8th International Conference on Applied System Innovation (ICASI)*, pp. 58–63. IEEE (2022).

Soyka, Florian, Kokkinara, Elena, Leyrer, Markus, Buelthoff, Heinrich, Slater, Mel, and Mohler, Betty. "Turbulent motions cannot shake VR." In *2015 IEEE Virtual Reality (VR)*, pp. 33–40. IEEE (2015).

Vora, Jeenal, Nair, Santosh, Gramopadhye, Anand K., Duchowski, Andrew T., Melloy, Brian J., and Kanki, Barbara. "Using virtual reality technology for aircraft visual inspection training: Presence and comparison studies." *Applied Ergonomics* 33, no. 6 (2002): 559–570.

Williamson, J. R., McGill, M. and Outram, Khari. "Planevr: Social acceptability of virtual reality for aeroplane passengers." In *Proceedings of the 2019 CHI Conference on Human Factors in Computing Systems*, pp. 1–14. (2019).

Wu, Wen-Chung, and Van-Hoan, Vu. "Application of virtual reality method in aircraft maintenance service—Taking Dornier 228 as an example." *Applied Sciences* 12, no. 14 (2022): 7283.

9 AI, Robotics, and Autonomous Systems
Assessing AI Applications in Aviation Industry

Büşra Yiğitol
Necmettin Erbakan University, Konya, Türkiye

9.1 INTRODUCTION

Technological development, also known as technological progress, refers to the continuous development and improvement of technologies over time. These improvements can be incremental where existing technologies are improved, or revolutionary where completely new technologies are introduced. Technological development plays a crucial role in shaping various aspects of human life (Watt, 2023). Technological developments have a significant impact on society, economy, and general well-being (OECD, 2000; Niggli & Rutzer, 2023). Especially in terms of production, technological developments increase the efficiency and productivity of various processes. For example, automation and robotics are revolutionizing manufacturing by streamlining production lines and reducing human labour (China, 2023). From a communications standpoint, advances in communication technologies such as the internet and mobile devices are revolutionizing the way people connect globally (Rogers, 2019). Thanks to these technologies, in addition to instant communication, they facilitate access to information and international cooperation. Thanks to the developments in the field of information technologies, business processes have gained a different dimension with faster processors, increased storage capacities, and more efficient software (CEPAL, 2021). In terms of sectors, the advantages of technological developments can be seen. Technological advances in the healthcare industry have led to better medical treatments, diagnosis, and patient care. Medical devices, advanced imaging techniques, and personalized medicine are some examples of how technology is transforming healthcare (Joshua et al., 2022; Paul et al., 2023). From the perspective of the education sector, online learning platforms are transforming education by providing interactive educational tools and access to vast amounts of information and resources (Zhang & Aslan, 2021; Haleem et al., 2022). Significant technological developments are seen in the field of transportation, from the development of automobiles to the emergence of electric and autonomous vehicles. These innovations improve safety, reduce emissions, and increase transport efficiency (Guerrero-Ibáñez et al., 2018).

While technological developments make life easier, they also reduce the negative impact of humans on the environment (Boserup, 1981). Technological advances address environmental challenges by developing environmentally friendly solutions and reducing the environmental footprint of industries and daily activities (Javaid et al., 2022). In addition, the search for sustainable energy sources has led to advances in renewable energy technologies such as solar, wind, and hydroelectric power. These technologies reduce dependency on fossil fuels and combat climate change (Panwar et al., 2011). As a matter of fact, it should not be forgotten that technological developments have advantages and disadvantages. In addition to the advantages provided, technological developments also bring challenges such as ethical concerns, cybersecurity threats, and possible dismissals due to automation (Cyert & Mowery, 1987; Aubert-Tarby et al., 2018). However, the important thing at this point is to maintain balance. Striking a balance between embracing technological progress and addressing its potential disadvantages is a crucial aspect of managing its impact on society.

In general, technological developments aim to improve the quality of life by facilitating tasks in both daily and business life, improving access to goods and services, and solving societal challenges. With the latest Industry 4.0 technologies, this process is gaining even more speed. With the introduction of Industry 4.0 solutions in many sectors, quality of life has increased. One of these sectors is the aviation industry. In this study, Industry 4.0 technologies were evaluated in terms of the aviation sector. In this context, in the first part of the study, the concept of Industry 4.0 and its technologies are discussed. In the second part of the study, the use of Industry 4.0 technologies in various sectors is evaluated, and the relationship between the aviation industry and technology was examined. In the third part of the study, a literature review is given, and in the final part, Industry 4.0 technologies in the aviation sector are examined.

9.2 INDUSTRY 4.0 AND TECHNOLOGIES

In the 21st-century, businesses are witnessing the digital transformation referred to as Industry 4.0. Industry 4.0 refers to the use of digital processes in the way products are produced and delivered in the 21st-century (Figure 9.1). The concept represents a transformative approach towards manufacturing and industry in the business world (Hofmann & Rüsch, 2017). It refers to a process in which smart materials and smart machines communicate with each other, interact with the environment, and make decisions with minimal human participation through interconnected computers (Ghobakhloo, 2020).

Industry 4.0 defines a revolution in the traditional industry. The main theme of this revolution is the integration of digital technologies, automation, and data exchange. This represents a significant change in the way manufacturing and other industries operate, enabling more efficient and smarter processes. Digital transformation processes are causing a significant change in the way manufacturing and other industries operate. Transformation processes provide many benefits such as increased productivity, reduced costs, improved product quality, improved resource management, and more sustainable production practices (Santos et al., 2018; Enyoghasi & Badurdeen, 2021). In addition to these benefits, Industry 4.0 also presents some difficulties,

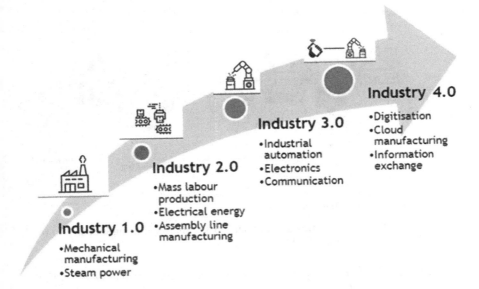

FIGURE 9.1   Industrial revolutions.

(Source: Prinsloo, J., Vosloo, J. C., & Mathews, E. H. (2019). Towards Industry 4.0: A roadmap for the South African heavy industry sector. *South African Journal of Industrial Engineering*, 30(3), 174–186.)

including implementation, the need to increase the skills of the workforce, addressing cybersecurity risks, and ensuring interoperability between different technologies and systems (Mohamed, 2018; Alani & Alloghani, 2019).

Industry 4.0 represents digital processes in data transfer with automation systems in manufacturing processes and defines the state of digital technologies shaping industries: a breakthrough in how industries integrate advanced technologies into their systems. The main purpose of the fourth industrial revolution is to create smart factories that are highly efficient, adaptable to various variability, and allow customization in line with customer demands (Venkateswaran, 2020). Integrating with the latest technology and adapting these technologies to the systems helps to capture the changing market understanding of the 21st century, while helping to maintain its competitive position. At the same time, it paves the way for future production and progress. In this context, it is possible to summarize the technologies covered (Figure 9.2) and the features offered by Industry 4.0 as follows (Xia et al., 2012; Purcell, 2013; Gibson et al., 2014; Li et al., 2017; Rao et al., 2018; Frank et al., 2019; Boston Consulting Group, 2019; Ratnasingam et al., 2019; Butt, 2020):

- *Internet of Things (IoT): The Internet of Things (IoT), refers to all systems capable of transferring data from and to humans over a network without the need for a network, provided by interrelated computing devices, mechanical and digital machines, objects, animals, or unique identifiers (UIDs). IOT*

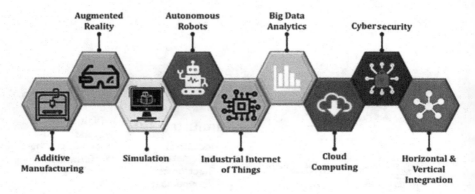

FIGURE 9.2   Industry 4.0 technologies.

(Source: Butt, J. (2020). A strategic roadmap for the manufacturing industry to implement industry 4.0. *Designs***,** *4***(2), 11.)**

> enables real-time monitoring and control, providing seamless communication and data sharing between components.
>
> - *Big Data and Analytics: Big data is a concept that defines heterogeneous data in different volumes that cannot be processed using traditional database techniques and consists of various digital contents. Digital technologies produce large volumes of data about machines and processes in the industry. Big data analytical processes are used to analyze and optimize this large amount of data, to predict the maintenance needs of the processes, and to make the appropriate decisions accordingly.*
> - *Cloud computing: Cloud computing is the presentation of all systems such as servers, storage, databases, networks, software, analytics, and machines in the cloud; i.e., on the internet. It represents the infrastructure that functions for storing, processing, and preparing large volumes of data coming from machines and processes in the industrial field. Cloud computing technologies make it possible to share data among stakeholders without any problems. While this technology provides a flexible and measurable function, it also supports businesses in adapting to changing demands.*
> - *Automation and Robotics: Robotics technology refers to the field of production of devices that can perform specified tasks, consist of electronic and mechanical units, have sensing ability, and can be programmed. Robot technologies play an important role in digital transformation processes. Robot technologies play a more active role in higher consistency, precision, and repetitive tasks. Robotic systems can reduce the need for human intervention in repetitive or dangerous processes.*
> - *Artificial Intelligence (AI) and Machine Learning (ML): Artificial intelligence and machine learning technologies tend to make predictions with the help of data and accordingly increase the future performance of the system. The purpose of artificial intelligence is to simulate natural intelligence in solving complex problems. On the other hand, machine learning is designed to learn*

from data related to a specific task. These technologies serve as a support for optimizing production processes, thereby increasing efficiency and quality.

- *Additive manufacturing technologies allow the production of desired parts in a single process without the need for an additional production process. The production of complex prototypes is facilitated by additive manufacturing technologies. The production of customized products and sub-industry or spare parts can be easily produced with additive manufacturing technologies. Thus, in addition to saving time, waste of material can be prevented.*
- *Cybersecurity is a critical issue with the development of technology and increasing connections. In its most basic sense, cybersecurity is the practice of protecting electronic systems, mobile devices, servers, computers, networks, and data from malicious attacks. With digital transformation and Industry 4.0, industrial systems, the data produced by these systems and the networks connecting the systems are potentially threatened, so protecting these systems is one of the priority areas.*
- *Cyber-physical systems are structures that include communication and coordination between the cyber world and the physical world. The fourth industrial revolution foresees the integration of digital processes. Sensors allow physical systems and machines to be connected to each other, to exchange data, and even to make decisions independent of humans. These systems are essential parts of communication between the physical and the cyber worlds.*
- *Augmented Reality (AR) and Virtual Reality (VR): In Industry 4.0, AR and VR technologies support maintenance, training, and product design processes. They provide immersive and interactive experiences that increase employee productivity and reduce errors.*
- *Horizontal and Vertical Integration: Industry 4.0 promotes integration at different stages of the production process. Horizontally, it connects various components and processes along the value chain. Vertically, it connects different levels of the organization from the production floor to top management for better coordination and decision making.*
- *Simulation and Digital Twins: Manufacturers can create virtual copies of their physical assets and processes, known as digital twins. These simulations help test and optimize processes before they are implemented in the real world, thereby reducing the risk of errors and downtime.*

Industry 4.0 provides many advantages for both society and organizations with the technologies it offers. These advantages include (Ibarra et al., 2018; Mohamed, 2018; Müller et al., 2018; Koilo, 2019; Pietrewicz, 2019; Javaid et al., 2020; Ghobakhloo, 2020; Mckinsey, 2022, Beier et al., 2022; Ding et al., 2023):

- Increased Efficiency: Industry 4.0 technologies optimize production processes, making them more efficient and reducing waste. This leads to lower resource consumption and a more sustainable use of materials. Automation and data-driven decision making lead to improved productivity, enabling organizations to produce more with less.

- Enhanced Safety: The use of robot technology and automation in hazardous tasks improves workplace safety by reducing human exposure to hazardous environments and repetitive tasks.
- Customization and Personalization: Industry 4.0 enables mass customization, resulting in greater customer satisfaction by allowing products to be tailored to individual preferences and needs.
- Improved health and wellbeing: In the healthcare industry, IoT devices and AI-driven analytics allow remote monitoring of patients and early detection of health problems, leading to better healthcare outcomes and improved wellbeing.
- Sustainability: together with digital technologies, the fourth industrial revolution embraces sustainability. Digitization contributes to the creation of sustainable business practices. It influences the adoption of future-oriented approaches in many aspects, especially in the environmental, and indirectly focuses on the solution of environmental problems with the reduction of resource waste and effective resource management.
- Real-Time Data and Analytics: Changing competition and market conditions require businesses to follow trends instantly. Therefore, rapid access to up-to-date and accurate information is a critical issue. To make informed decisions quickly, it is necessary to access the data needed at any time. Today, digital technologies enable instant access.
- Ease of access: Advanced technologies also make accessibility more possible. Currently, it is possible to see examples of how effectively technology can be used in information transfer among stakeholders in many fields. It is widely used to make information accessible to people from anywhere in the world, especially in the education sector, and to support the education and skill development processes of individuals.
- Growth and employment: Although there is a perception that new technologies will have negative effects on employment, they actually trigger the formation of new business lines. On the other hand, new business areas provide an increase in employment in new areas. This contributes to the creation of new business opportunities. new business lines and accordingly expanding employment are one of the building blocks that will stimulate economic growth.
- Innovation: Digital technologies and Industry 4.0 are important for transforming new ideas into value-creating outputs. Technological developments prepare the ground for innovation.
- Competitive Advantage: The competitive approach of the 21st-century is undergoing a difficult process. Changing market conditions make it necessary for businesses to be more flexible, fast and in harmony with changing conditions. Digital technologies play a supporting role in helping businesses gain agility that will provide competitive advantage.
- Cost Advantage: Digital processes record data about systems instantly. The data obtained play an active role in predictive maintenance planning. In this way, it is possible to detect possible malfunctions in advance. The reduction of downtime and the sustainability of production have noticeable effects

on cost. This is positive for businesses because it also provides resource savings and optimization.

- Global Cooperation: Digital technologies enable the creation of networks of connections around the world. These connections pave the way for the development of a collaborative approach on a global scale. In addition, businesses get the opportunity to work with transnational partnerships, suppliers and customers.
- Quality and Customer Satisfaction: Increasing product quality is related to following technological developments. This technology allows the production of products with fewer errors. In this context, digital technologies allow the preservation of quality in the manufacturing processes, from product planning to the final product. In fact, it indirectly contributes to customer satisfaction and loyalty.
- Supply Chain: Digital technologies increase visibility, forecast accuracy, real-time information transmission speed, process optimization, and communication power throughout the supply chain. In this context, it provides support for enterprises to make accurate logistics, inventory management, and future demand forecasts.

Overall, Industry 4.0 provides transformative benefits for both society and organizations. It encourages businesses to innovate and thrive in the rapidly evolving digital landscape, while promoting a more connected, efficient, and sustainable world.

9.3 INDUSTRY 4.0 IN SECTORS

Industry 4.0 has been implemented in various sectors to increase efficiency, productivity, and innovation. In the manufacturing industry, IoT sensors are used to predict when machines may require maintenance or replacement, and to monitor equipment health in real time (Ayvaz & Alpay, 2021). This helps reduce unplanned downtime and optimizes maintenance schedules (Kanawaday & Sane, 2017). It also creates digital copies of physical assets and production lines to simulate and optimize processes, resulting in better resource utilization and product quality. 3D printing can be used to create prototypes, custom parts, and complex designs quickly and cost-effectively (Berman, 2012). In addition, in some manufacturing sectors, human-robot collaboration is used to increase production efficiency, especially in tasks that require precision and speed.

In the healthcare industry, IoT devices and wearables are used to remotely monitor patients' vital signs and health status (Iqbal et al., 2021). Such remote monitoring systems allow timely intervention and reduce the length of hospital stays. With big data analytics and artificial intelligence technologies, personalized processes can be used to analyze patient data and genetics and can be made into special treatment plans based on individual characteristics. Artificial intelligence-supported algorithms help interpret medical images, and provide faster and more accurate diagnoses (Bag et al., 2023).

In the transportation and logistics industries, the emergence of new logistics fields such as robots and software using artificial intelligence and drones in transportation direct companies to new software. The logistics sector is dynamic sector that works based on real-time data. Various software can be used to monitor vehicles in real time and to monitor vehicle performance. One of these software is IoT technology. IoT-enabled sensors provide better fleet management and optimized routing for logistic companies (Charoenporn, 2018). Industry 4.0, which is a part of the digital world, carries the logistics industry to a different dimension with supports such as warehouse areas, autonomous transport systems, and smart walking paths. In logistics processes, digital technologies are used to automate warehousing operations, including inventory management and order fulfilment (Akkaya & Kaya, 2019; Buntak et al., 2019).

Industry 4.0 technologies have been frequently used in the field of energy. Smart grids use IoT and data analytics to optimize energy distribution and consumption, resulting in more efficient energy use and reduced waste. IoT sensors are used in renewable energy systems (solar panels, wind turbines) to monitor their performance and maximize energy production (Borowski, 2021).

In the retail industry, RFID technology and IoT devices are used for real-time tracking of inventory levels, enabling retailers to maintain optimum stock levels and reduce stock outs (Unhelkar et al., 2022). Artificial intelligence and big data analytics are used to analyze customer preferences and behaviour, allowing retailers to offer personalized recommendations and targeted marketing.

In addition to the manufacturing, transportation, and health sectors, Industry 4.0 technologies are used in the field of agriculture to improve processes. IoT devices, drones, and AI-assisted analytics are used to precisely monitor crops, soil conditions, and weather, which supports optimized irrigation and fertilization practices (Zambon et al., 2019; Liu et al., 2020). Again, IoT sensors can be used to monitor the health and wellbeing of animals (Farooq et al., 2019; Simitzis et al., 2021). Thus, an early detection of diseases is possible and general herd management is improved.

As the examples mentioned above can be increased, all show how Industry 4.0 technologies are transforming various industries, streamlining operations, and creating new opportunities for growth and innovation.

9.4 AVIATION INDUSTRY AND TECHNOLOGY

The aviation industry is a vital component of modern transportation and global connectivity. It encompasses various activities related to the design, manufacture, operation and maintenance of aircraft and the infrastructure and services that support air travel (AviationHunt, 2023). It plays a pivotal role in both commercial and military applications and shapes the worldwide travel of people and goods (NCI, 2022).

The aviation industry has faced numerous challenges over the years, including economic downturns, fuel price fluctuations, environmental concerns, and the impact of global events such as pandemics (Heiets & Yibing, 2021). However, remarkable growth and innovation is making air travel more accessible, efficient, and safer than ever before. In particular, the R&D activities carried out this growth further beyond its borders (AIAA, 2023). Research and development is continually invested in improving safety, efficiency, and environmental sustainability (FAA, 2022). In fact, developments in aerodynamics, materials science, propulsion technologies, and especially the integration of digital solutions reveal the results of these investments.

In terms of technology and digital solutions, the aviation industry is very open to development. There is a symbiotic relationship between technology and the aviation industry. Technology has been the driving force behind the progress of the aviation industry (Wittmer & Vespermann, 2011). Technological innovations are constantly shaping the aviation industry, enabling it to reach remarkable milestones in aerospace research. On the contrary, the demands and challenges of the aviation industry also trigger technological development. Today, it is possible to see common usage examples of the solutions offered by technology in many fields, from aircraft design to propulsion systems, from avionics and flight systems to air traffic management and even space exploration studies. In aircraft design, Innovations in materials, aerodynamics, and computer-aided design (CAD) are enabling the construction of more efficient and capable aircraft. Lightweight materials such as carbon composites and aluminium alloys make aeroplanes stronger and more fuel-efficient (Spencer, 2022). Advanced CAD software and simulation tools allow engineers to virtually model and test aircraft designs, reducing the need for physical prototypes and speeding up R&D processes. With technological developments in the aircraft industry, more powerful and fuel-efficient engines can be developed, and thanks to gas turbine engines such as turbofans and turboprops, faster and longer flights can be made with reduced fuel consumption. These developments are revolutionary in commercial aviation. In recent years, efforts to develop electric and hybrid-electric propulsion systems have continued to increase efficiency and reduce environmental impact (Murugan, 2023). In summary, technology and the aerospace industry are deeply interconnected, and each enables the other to advance. As technology continues to advance, it is also likely to make further breakthroughs in aviation, leading to safer, more efficient, and sustainable air travel and space exploration.

9.5 LITERATURE REVIEW

The number of studies examining Industry 4.0 technologies in the aviation sector is increasing daily. This section summarizes the review of the available literature, which includes both academic articles and industry reports.

Ramalingam et al. (2017) aimed to determine the characteristics of various general IoT technologies that are not related to a specific industry or field, and to investigate the corresponding IoT features in the aviation field. The methodology presented reveals the IoT features of aerospace systems businesses and their associated impact on aviation systems.

Rodrigues et al. (2022) proposed an IoT-based solution to focus on the challenging field of the aerospace industry in their work, improving the quality of monitoring data that helps existing work, production, and assembly processes. The solution proposed in this study contributes to better management of production resources and enables shorter production times and faster and more informed feedback.

Badea et al. (2018) included big data applications in the aviation industry, the main advantages of big data, and a case study. In the results of the study, benefits such as low maintenance costs and improved aircraft availability through optimization of the maintenance programme, low inventory requirements for parts that need to be replaced through integrated distribution chain and planning, fuel economy, and optimizing flight plans are listed as the advantages of using big data in aviation.

Bogue (2018) provides details of the increasing use of robots in the aviation industry. The results of the study drew attention to the increasing use of robots in various applications in the aviation industry. In the results of the study, the benefits of robots to the aviation industry are listed as low costs, manpower, and time scales, improved quality, and new production capabilities.

Kumar and Srinivas (2019) focus on robot applications in the aerospace industry. Examples of usage areas of robots in processes such as defence industry, aircraft production processes, and airport passenger services are presented.

Adamopoulou and Daskalakis (2023) presented a literature review of leading articles on the application of big data in aviation. A total of 67 publications emphasizing the sources, uses, and benefits of big data were analyzed, and titles such as aviation technology and aviation management, UAV-supported applications, military aviation applications, health/environment-related applications, and applications in space technology were used to categorize the publications. the results of the study draw attention to the benefits of big data in aviation, health, environment, humanitarian operations, network communications, etc.

Singh et al. (2016) proposed a baggage tracking and handling system design using smart RFID tags and cloud server-based IoT to provide a better and safer system to passengers. With the proposed technology, baggage loss can be prevented by facilitating step tracking, the real-time location of baggage can be tracked using IoT, and data can be stored via cloud technologies.

Hussain et al. (2019) examined the technologies offered by Industry 4.0 in the fields of aerospace and defence industry. In this study, the main focus areas of aerospace and Defence 4.0, the digital transformation roadmap, and aerospace and Defence 4.0 concepts were examined.

In a report prepared by the European Commission (2017), the application areas of IoT technologies, one of the Industry 4.0 technologies in aviation, were examined. While IoT technologies are described as a radical transformation, information about the opportunities of IoT solutions is also compiled, and its impact on the value chain and business models is stated.

9.6   USE OF INDUSTRY 4.0 IN AVIATION

In this study, Industry 4.0 technologies are examined in terms of the aviation industry. In this study, a compilation is presented on secondary data obtained by scanning many domestic and foreign internet resources and articles.

9.6.1   FINDINGS OF THE RESEARCH

Changing market needs and customer expectations are fundamentally changing the way aircraft are designed and manufactured. To build aircraft that meet the highest quality and performance standards, future-oriented, smart, and digital solutions must be used in the entire industry, especially in production systems. In this respect, various Industry 4.0 technologies are applied in today's aviation industry to increase efficiency, productivity, safety, and sustainability. In light of the result, the areas of benefiting from Industry 4.0 technologies are stated below.

9.6.1.1 Internet of Things (IoT)

Aviation is one of the most complex and challenging industries in the world. Considering the difficulty of tracking many aircraft at the same time, it becomes even more noticeable why it is a difficult sector. Tracking and controlling these planes, which move to almost every part of the world, is an exceedingly difficult process. In particular, the increasing need for efficiency emphasizes the necessity of providing data for systems in the aviation sector. Technologies that allow data transfer between sensors and computers, such as the Internet of Things, are seen in the aviation sector as well as in many other sectors.

IoT plays an important role in aviation by connecting various devices and systems to collect real-time data (Airbus, 2019). On board, sensors are used to monitor equipment health, fuel levels, and operational parameters (Turbide, 2017). These sensors on the aircraft act as intermediaries for data collection, and can help detect potential problems before they affect the passengers or pilots during the flight (Brownlow, 2022). Data collected through sensors are transmitted to ground stations for analysis, allowing predictive maintenance and reducing downtime. The monitoring and planning area mainly focuses on how airlines use IoT systems in aviation to track where each aircraft is going at all stages of its journey, from the departure gate to the destination. IoT applications in the aviation industry increase their performance by facilitating maintenance, fleet management, and security processes (Lee & Lee, 2015). The aviation industry is highly regulated, focusing on safety, quality, and compliance.

IoT applications improve flight security in various ways (Singh et al., 2016; Turbide, 2017; Airbus, 2019; Mariani et al., 2019; Stewart, 2021; Brownlow, 2022; IOTSWC, 2023):

- The sensors enable diagnosis to be made in real time. It allows real-time monitoring of aircraft systems so that problems can be detected before they cause safety risks to passengers or crew members. Equipment failures can be detected quickly. By tracking aircraft performance, it becomes easier to understand the areas that need to be addressed, streamlining the maintenance process with targeted treatment, minimizing repair times, and maximizing safety.
- Aeroplanes are complex structures with different components. It is important that these components are equipped with sensors to measure certain settings and that the data produced from these sensors is collected and controlled by a central system. First, the aircraft's communication with ground controls allows for the necessary follow-up and diagnosis of problems in terms of efficiency and safety. IoT technology enables real-time data acquisition. Real-time data have a positive effect on aircraft uptime, safety, and maintenance processes. Thus, high value-added feedback can be provided to both the airline company and the customer. This is exactly what Virgin Atlantic does. Boeing provide a good example of implementation with its 787 fleet.
- The possibility of unexpected delays can be minimized by monitoring fuel levels and engine performance. This allows airlines to use their time as efficiently as possible.

- IoT technology is used to anticipate potential hazards and flight-related risks. The collected data can be used for post-flight analyzes, and alternatives can be produced for the flight route. In some cases, emergency interventions may be necessary. IoT technologies serve as support in emergency route changes to ensure the safety of flight crew and passengers in bad weather conditions.
- IoT can be used to identify the most efficient routes. Combined with optimizing fuel consumption, this can significantly reduce airline costs. Using the most efficient routes while optimizing fuel consumption results in lower costs and therefore maximizing profits. AirAsia has already implemented a system with these features that will save an estimated $30–50 million over five years.
- In emergencies that require the immediate attention of air traffic control personnel, accurate location information should always be available so that they can act quickly without delay. IoT systems provides the necessary support to the flight crew in this regard.
- Elements such as climate control in the cabin can be automatically controlled and optimized by deploying temperature sensors strategically placed throughout the cabin. This includes temperature, humidity, lighting, and sounds. On the other hand, the data flow should help, for example, provide more information to those waiting for their flight at the airport and further automate the check-in process.
- Baggage can be tracked and shared with airlines and passengers to identify where their baggage is at any given time. Thanks to RFID tags, a technology implemented by Delta Air Lines, being unable to find luggage may soon be a thing of the past for airline customers. The same technology can be used by shipping companies to track goods. In addition, tracking data are sent to passengers or shipping companies, allowing them to track where their belongings are at any time and practically from their smartphones.

In summary, IoT technologies can increase security and efficiency at every stage of the process in many industries, including aviation. It acts as a guide for managers to make correct and accurate decisions about maintenance and planning. Real-time data collection with sensor technologies in the fleets of airline companies makes operational efficiency and optimization possible. Thus, cost savings can be achieved. Such technologies make important contributions to prevent unexpected flight delays and delays and ensure fuel efficiency.

9.6.1.2   Big Data Analytics

Big Data represents next-generation technologies and architectures designed to extract value from a wide variety of large data volumes, allowing it to be processed and analyzed in real time (Alexandru & Coardoş, 2017). The aerospace industry deals with large volumes of data generated from flight operations, maintenance, and other processes. A clear example of the impact of Big Data in the aviation industry is today's "digital aeroplane", which can collect up to 300,000 parameters depending on flight time and aeroplane type (Badea et al., 2018). In simple and practical terms, an average Boeing 737, a commercial twin-engine aeroplane, produces 20 terabytes of information per engine per hour during a six-hour flight. It can be said that by multiplying 20 terabytes of information from a single engine with 6 h of flight, 240 terabytes of data will be obtained from the engines. Each aircraft has its own unique number, flight departure and destination, number of passengers, route, and terabytes of technical data on each part and component that keeps the aircraft in the air. This is the information an aircraft generates during a flight. When we calculate this data size by considering the number of aeroplanes in the world and the number of flight days, a huge data flow emerges (Badea et al., 2018). At this point, it is further understood why big data and analytics applications in the aviation industry are important.

With big data, data on the status and operational performance of aviation operations are recorded and reflected, which helps to ensure aviation traffic safety and optimize aviation operations performance (Dou, 2020). Aviation analytics systems use all this information to analyze fuel efficiency, passenger and cargo weights, and weather conditions with a view to optimize safety and energy consumption. Big data analytics processes the collected data to identify patterns, optimize flight routes, predict potential failures, and improve overall performance. The role of big data in route planning and air traffic management will become increasingly prominent with the increasing number of aviation vehicles and shrinking airline resources.

Big Data simplifies and streamlines critical processes in aviation. These processes are:

- Congestion management and traffic control (Congestion management and traffic control): Thanks to Big Data analytics, Google Maps can now tell you the least traffic-prone route to any destination.
- Route planning: Different routes can be compared in terms of user needs, fuel consumption, and other factors to maximize efficiency.
- Traffic safety: Real-time processing and predictive analytics are used to identify accident-prone areas.

Thanks to big data, aviation companies can make more informed decisions and predict different scenarios.

9.6.1.3 Cloud Computing

Cloud technologies have become a high priority in the aviation industry. There are many cloud-based solutions currently used in the industry. This technology is frequently used, especially in predictive maintenance and condition monitoring processes of aircraft. Predictive maintenance requires the use of machine learning-driven algorithms to predict potential problems and preventive measures. This maintenance procedure reduces the need for unscheduled maintenance and associated downtime. Condition monitoring focuses on monitoring the real-time health of an aircraft and its components. This ensures accurate diagnosis and proactive intervention. When used in conjunction with predictive maintenance, it can help reduce technical difficulties, increase safety, and lower operational costs.

Cloud computing technologies also play an active role in forecasting weather events. It is stated in the studies that cloud computing technologies are used in real-time analytics that help predict weather conditions and air traffic. It also highlights that it can be used to simulate aircraft parts and test the aircraft device as a whole (Gupta & Rathore, 2012). With an aviation cloud that helps to monitor air masses in real time, optimized routes can be provided and customers can be well informed about weather conditions and ticket booking before the plane takes off (ITU, 2014; Vagdevi and Guruprasad, 2015).

The aviation industry has also started using cloud-based solutions for asset, personnel, and cargo tracking. This technology facilitates real-time monitoring and provides better monitoring and control (Lohse, 2023). In addition, the use of facial recognition systems is becoming more common in airports for better security. All these cloud-based applications increase security, reduce the cost of operations, and increase efficiency. Thanks to technological infrastructure, the aviation industry can manage its operations more effectively and provide a safer flight experience for everyone, with the promise of greater savings.

Large volumes of data are frequently used in the operations and management of the aviation industry. With the advent of cloud computing, aerospace companies can now analyze vast amounts of data sent to them by aeroplanes and satellites in real time. Using this knowledge can result in cost savings and increased productivity and security. Cloud technology can be used to protect highly sensitive data, such as those used by the military and government (Vagdevi & Guruprasad, 2015).

Businesses in the aerospace industry can also avoid investing in the entire infrastructure by using cloud computing and only paying for the services they use. Cloud technologies allow easy rebuilding of any aircraft component, rather than creating a real prototype (Delvadiya, 2023). Engineers and designers can create and test aeroplanes remotely using cloud-based collaboration technology. Therefore, cloud solutions can help the aviation industry create plans based on cloud-assisted data analysis. These technologies not only speed up the planning and testing process, but also save time and money compared with previous techniques.

9.6.1.4 Additive Manufacturing (3D Printing)

Aviation parts are not only extraordinarily complex, but must also be structurally sound and meet the highest quality assurance standards of almost every industry (Vishnuraj, 2022). These and similar challenges push technological developments in the aviation industry even further. To reduce costs and overcome traditional manufacturing challenges, many aerospace companies are taking advantage of the possibilities offered by technology by abandoning traditional manufacturing processes to efficiently manufacture the complex parts they need (Spatial Team, 2021). Additive manufacturing (3D Printing) is one of these areas. Additive manufacturing enables complex engine parts, structural components, and spare parts to be produced at lower weight and significantly reduces lifecycle costs (Oerlikon, 2023). Attachment manufacturing technology can be used not only in the production of complex parts, but also for weight and flow optimization, noise reduction, net part replacement, and part count reduction in aircraft applications such as brackets, ducts, and seat belt buckles (Sheng, 2022).

The benefits of additive manufacturing to the aviation industry can be listed as its positive effect on storage costs, as it facilitates the production of geometric parts, provides more efficient prototyping and cost advantage, facilitates the production of light parts, and provides the opportunity to produce when needed. The aviation industry uses equipment and tools that carry highly complex parts. From helicopter mechanisms to turbine engines, aerospace components sometimes require highly complex geometric structures in very tight spaces. Instead of creating small, complex parts separately and then joining them together as a whole, three-dimensional models of the entire structure can be created using CAD data (Spatial Team, 2021; Spiegel, 2022). The 3D printer can create a single seamless part containing all the complex interior dimensions and geometries without the need for assembly (Duda & Raghavan, 2016). With additive manufacturing, it is possible to produce using many aircraft materials. Thus, it is possible to create complex parts while reducing lead times (General Electrics, 2023). Of course, additive manufacturing has some disadvantages and its advantages. In particular, the limited component size and high initial costs are among the main disadvantages. In addition, some difficulties may arise in ensuring production according to the specifications. In aviation, it is crucial that a part is manufactured according to specifications. In conventional production, the specification process has already been established. However, with additive manufacturing, the specification process has not been established, so it is not how to ensure that a manufactured part is produced to specifications (Spatial Team, 2021). Another advantage it provides regarding production is more efficient prototyping. Product prototypes can be easily made in the field of aviation with the possibility of three-dimensional and modelling. Engineers and designers use 3D modelling software to create detailed representations of aircraft and spacecraft components. These models enable them to visualize designs, run virtual tests, and optimize performance before building physical prototypes. 3D printing technology enables rapid prototyping, cost-effective manufacturing, and the ability to create complex geometries that would be challenging or impossible with traditional manufacturing methods. Aerospace companies can accelerate their time to market and stay ahead of the competition with the ability to build and test prototypes faster (Sheng, 2022). Boeing uses 3D printing to manufacture plastic interior parts. Parts made of Ultem and nylon are mainly used to make prototypes and test coupons (Joshi & Sheikh, 2015). In addition, 3D technology is used to manufacture the tools necessary to produce composite parts (Fetters-Walp, 2011). Additive manufacturing can reduce not only prototyping time but also cost. Because materials are added and not removed in additive manufacturing, it can greatly reduce material waste and help manufacturers save on production costs (Joshi & Sheikh, 2015). Initial costs are associated with the establishment of additive manufacturing processes. However, in the long run, the cost savings outweigh these upfront expenses.

9.6.1.5 Artificial Intelligence (AI)

Artificial intelligence applications play an effective role in providing solutions to complex engineering and structural problems where traditional logical algorithms do not work properly (Das et al., 2021). Today, when we look at the industries, it can be seen that the uses of artificial intelligence are divided into two. The first is analytical artificial intelligence, which is a type of system that can predict when a machine will fail, detect credit card fraud, or suggest the next book to buy on

Amazon. Operational AI, the second class of AI applications, actually does something in the physical world. AI applications can manage a factory process, fly an aeroplane, drive a vehicle, or act on predicted events (DXC, 2023).

From the viewpoint of the aviation industry, it is seen that artificial intelligence solutions are becoming widespread in the sector. Artificial intelligence-supported systems can perform tasks such as autonomous take-off, landing, navigation, and obstacle avoidance. Even today, various applications such as surveillance, cargo delivery, and planetary exploration can be carried out using technologies such as unmanned aerial vehicles. When we look at artificial intelligence applications in aviation, it can be seen that it is used to optimize flight routes, reduce fuel consumption, and improve airspace management. Artificial intelligence is used especially in structural damage detection. Among the many engineering problems, early-stage structural damage assessment and structural health monitoring are major headaches for structural designers and aerospace engineers. Artificial intelligence techniques, such as Neural Networks (NN), Genetic Algorithms (GA), fuzzy logic, Adaptive Neuro Fuzzy Inference System (ANFIS) can mostly be used for error prediction in aircraft and spacecraft structures (Mujica et al., 2008). AI technology is used to optimize flight routes, reduce fuel consumption, and improve airspace management. AI-powered systems can improve safety and efficiency in air travel by analyzing weather data, air traffic patterns and airport conditions to provide real-time recommendations to pilots and air traffic controllers (Kell, 2023). AI-powered NLP systems can understand and process natural language commands and queries from pilots and air traffic controllers. This technology facilitates communication and reduces the risk of misinterpretation during critical operations (Yang & Huang, 2023).

Artificial intelligence applications can also be used for the maintenance management. Artificial intelligence and computer vision technologies enable the automatic visual inspection of aircraft components (Graham, 2023). Artificial intelligence algorithms are used to monitor the health of aircraft components in real time (Karaoğlu et al., 2023). By analyzing sensor data and historical maintenance records, AI can predict potential failures, suggest maintenance actions, and optimize maintenance programmes to reduce downtime and increase overall operational efficiency (Murugan, 2023). Artificial intelligence algorithms detect defects, cracks, and anomalies in aircraft structures, engines, and parts, making inspections easier and faster.

Artificial intelligence-supported systems can also be used in areas such as aircraft design and personnel training (Safi & Chung, 2023). It can be used to optimize aerodynamic designs, reduce drag, and increase fuel efficiency for new aircraft models. AI can analyze vast amounts of data and run simulations to determine the most efficient designs. In addition, it can be used in educational processes. AI-powered virtual assistants and training systems provide support to pilots and crew members during flight (Osborne et al., 2022). These systems can provide real-time information, aid decision making, and assist in training scenarios.

9.6.1.6 Automation and Robotics

With the advent of Industry 4.0, there has been an increase in the use of automated processes in global production. Although robot technologies are associated with the automotive and electronics industries, they are also widely used in the aviation industry today. Owing to their reliability, precision, and other superior features, aviation robots are one of the most widely used production solutions today (Dey, 2023). Robots play a vital role in aviation because of their precision, efficiency, and safety.

Robotic arms are used in spacecraft and satellites for tasks such as capturing and deploying payloads, servicing satellites, and assembling components in space. They offer the dexterity required for precision operations in microgravity environments (Fernandes & Ventura, 2020). Robotics applications can access hard-to-reach areas and perform routine inspections on aircraft structures, thereby reducing maintenance time and increasing safety. Robots are used in aerospace manufacturing for tasks such as assembling aircraft components, welding, and applying coatings (Anandan, 2016). It increases production efficiency and reduces the risk of human error. Robots are most frequently used in aerospace applications to drill holes in components (Owen-Hill, 2023). For example, thousands of holes may need to be drilled into an airframe, and because the job must be precise, robots are an excellent choice for consistent and fast results (Banks, 2021). Robots equipped with vision systems can locate where the robot should pierce the fuselage. Robots can improve the quality and efficiency of these processes by assisting with aircraft maintenance tasks such as painting, coating inspection, and surface treatments.

Robots can also be used for surveillance, reconnaissance, and tracking tasks in aviation. Unmanned robotic spacecraft and rovers are sent to explore other planets, moons, and celestial bodies (Sharma et al., 2021). These robots collect data, conduct experiments, and capture images to expand our understanding of space and search for signs of life. Drones can fly autonomously to collect data and images in real time, assist search and rescue operations, monitor airspace, and conduct environmental surveys (Jouav, 2023). These examples demonstrate how robotics plays an important role in aviation, enabling space exploration, improving aircraft production and maintenance, and increasing overall efficiency and safety in the aviation industry.

9.7   CASE STUDY: SAMPLE APPLICATIONS OF AI, ROBOTICS, AND AUTONOMOUS SYSTEMS IN AVIATION INDUSTRY

The aviation industry is a sector affected by the course of technological developments. Recently, the development of digital technologies has led to new developments in the field of aviation. The sector gains a different dimension, especially with the integration of AI, robots and automatic systems. It is possible to see the usage areas of AI, robots, and automatic systems in the field of aviation in the report published by EASA in 2023. EASA (2023) has published the new Artificial Intelligence Roadmap 2.0, which aims to develop the "human-centred approach to integrating artificial intelligence into aviation". The report emphasizes that artificial intelligence-based systems can provide new opportunities and focus more on competitiveness regarding problems such as the increase in air traffic volumes in the field of aviation, more stringent environmental standards, and the increasing complexity of systems. In this context, it is stated that artificial intelligence and machine learning-based systems bring tremendous potential for the development of new applications to replace traditional techniques regarding aircraft design and operation. In the report, deep learning (DL) brings with it a wide range of applications that can benefit aviation; It is also emphasized that it can open the door to solutions such as high-resolution camera-based traffic detection or virtual assistance to the pilot. In addition, the possible effects of digitalization in the fields of production and maintenance are mentioned.

Digital twins are developing in the aircraft manufacturing industry and Internet of Things technologies are becoming widespread. In this context, the report mentions processes such as the development of predictive maintenance, where large volumes of data are collected. For example, Airbus Aircraft Maintenance Analysis (Airman), used by more than a hundred customers, constantly monitors the situation, and provides instant information flow by transmitting malfunctions or warning messages to the ground control unit. This provides quick access to prioritized troubleshooting steps. Similarly, EASA's report also mentioned artificial intelligence applications in air traffic management.

The areas where AI provides support in the field of aviation, both in EASA's report and in many sources, can be summarized as follows:

- INTUIT project: The aim of the project is to improve our understanding of the trade-offs between Key Performance Areas (KPAs) (safety, environment, capacity, efficiency) in Air Traffic Management (ATM) and to develop new decision support tools for monitoring and managing ATM performance. The project is a SESAR exploratory research project exploring the potential of visual analytics and machine learning (SESAR, 2018).
- DART and COPTRA projects are developing an ML-based trajectory prediction capability to predict aircraft performance before or during flight (SESAR, 2015).
- Singapore ATM research institute has developed an artificial intelligence application that collects and shares information such as weather forecast and airport traffic in real time and in advance (Tan & Yang, 2023).
- BigData4ATM project explored how different passenger-centric geolocation data can be analyzed to identify patterns in passenger behaviour, door-to-door travel times and travel mode choices (BigData4ATM, 2016).
- The MALORCA project (Machine Learning of Speech Recognition Models for Controller Assistance) has designed a versatile, low-cost solution that adapts speech recognition tools for use at any airport (SESAR, 2016).
- SESAR's time-based separation (SESAR, 2023).

Artificial intelligence technologies are one of the areas that Airbus, a leading name in the field of aviation, attaches importance to. Airbus states that it uses artificial intelligence technologies in the fields of information extraction, computer vision, anomaly detection, speech assistance, decision making, and autonomous flight. For example, Airbus's ASTARTES project is an Artificial Intelligence (AI) project that aims to digitize human-level experience to support operators in tactical coordination tasks. Similarly, Airbus is enabling self-piloting commercial aircraft to take off and land using AI-based computer vision and machine learning technologies (Airbus, 2023).

Many projects, such as the projects listed above, offer usage areas of subjects such as artificial intelligence and machine learning in the field of aviation. It seems to be used in robotic processes as well as artificial intelligence and machine learning.

Many corporate companies such as Iris Automation Inc., The Boeing Company, Nvidia Corporation, Thales SA, Airbus SE, IBM, SITA, Lockheed Martin, General Electric and Intel Corporation are leaders in the field of artificial intelligence and robotics in the aerospace and defence market (Market Reports World 2023). NASA uses the Perseverance rover to explore the surface of Mars. Featuring special cameras and an artificial intelligence unit, this roving robot maps the entire terrain in its view and uses object detection to identify unique features. Nowadays, virtual assistants are also being developed to help and guide astronauts. Airbus' CIMON-2 is an example of this. CIMON-2 can talk, understand, and move with the astronaut. It also has the ability to act according to voice commands and perform assigned tasks. Moreover, it can show astronauts information and instructions for engineering tasks (Sajid, 2023).

It is also possible to see robots actively at airports. At Philadelphia International Airport, PHL Food & Shops is piloting a contactless ordering system that includes robotic food delivery. This robot's name is Gita. Standing 26 inches tall and capable of carrying up to 18 kg, Gita can navigate busy, pedestrian-filled concourses. It can also travel the equivalent of a 20-mile walk on a single four-hour charge. Similarly, robots are used for cleaning work. Heathrow Airport is using cleaning robots around airport terminals and lounges to disinfect areas using ultraviolet (UV) light, which has been proven to effectively kill harmful viruses and bacteria to provide a safe, secure environment for passengers (Youd, 2021). Another example of robot application at airports is in South Korea. Incheon International Airport Robot Assistant AIRSTAR is a second-generation robot assistant produced by LG. At the airport, he directs passengers to where they need to go, tries to answer all their questions and helps them take selfies. The robot speaks English, Chinese, Japanese, and Korean (Fraser, 2019). Similarly, Belfast International Airport also introduces robotic technology to its customers. Two robots named BellaBot and HolaBot support the service and cleaning work of the restaurant at the airport (Airport World, 2022). Robot technologies find use not only in airports but also in aircraft production and maintenance. VALERI project is one of them. VALERI stands for "Validation of Advanced, Collaborative Robotics for Industrial Applications". With the project, mobile and autonomous robots that will be used in the production of aircraft components and work side by side with humans are being developed. In the project, new tactile interaction methods will be tested (Saenz, 2015).

In aircraft manufacturing, Airbus (Toulouse, France) uses robotic processes developed by Loxin (Pamplona, Spain) to assist in the fabrication of fuselages for its best-selling A320 aircraft. Robots called Luise and Renate perform drilling and riveting operations on aircraft fuselages before assembly. Stating that it can drill almost 80% of the holes in the upper area of the aircraft and improve the ergonomic working environment, Airbus aims to increase aircraft production from 50 to 60 aircraft per month (Renner, 2019).

Northrop Grumman is developing processes that allow advanced automation capabilities to be used in conjunction with robots to produce the centre fuselage of the F-35 Lightning II. In this context, it produces many systems, including automation systems and the Communication, Navigation and Identification (iCNI) system for warplanes. The company pioneers automated production systems in the field of military aviation (McKinney, 2023).

9.8 CONCLUSION

Digital transformation is not limited to any industry. In this respect, Industry 4.0 trends have already shaped many sectors and created at source of competition. The aviation industry is also among these sectors. As Airbus states, changing market needs and customer expectations are fundamentally changing the way today's aircraft are designed and manufactured. To build aircraft that meet the highest standards of quality and performance, the industrial ecosystem must be future-oriented, intelligent, and digital. In this respect, digital technologies may help design parts that minimize the impact of fuel costs, even if they do not directly affect the fuel price. In this study, Industry 4.0 technologies were evaluated in terms of the aviation sector.

When evaluating Industry 4.0 technologies used in the aviation industry, the following points should be considered:

- Aircraft consists of many systems, each of which produces its own data. From this perspective, an aircraft produces large volumes of data. Cloud solutions are used in aviation to store these data and provide instant access.
- In addition, the Internet of Things (IoT) plays an important role in aviation by connecting various devices and systems to collect real-time data.
- With additive manufacturing technologies, the production of geometric parts is facilitated, more efficient prototyping is possible, cost advantage is provided, production of light parts is facilitated, and storage costs are reduced by providing the opportunity to produce when needed.
- While artificial intelligence technologies perform tasks such as autonomous take-off, landing, navigation, and obstacle avoidance in aircraft, they also enable various applications such as surveillance, cargo delivery, and planetary exploration with equipment such as unmanned aerial vehicles.
- Robotics assist in aircraft maintenance tasks such as airframe drilling, painting, coating inspection, and surface treatments. Thus, human error is brought closer to zero, and the quality and efficiency of the processes can be increased.
- Robots can also be used for surveillance, reconnaissance, and tracking missions.
- Artificial intelligence-supported systems can be used in areas such as aircraft design and personnel training.
- The aviation industry has started using cloud-based solutions for asset, personnel, and cargo tracking.
- Big data simplifies and streamlines critical processes in aviation, such as traffic control, route planning, and traffic safety.
- Aviation companies can make more informed decisions and predict different scenarios using big data.
- Cloud computing solutions are used in predictive maintenance and condition monitoring of aircraft.
- Cloud computing technologies also play an active role in forecasting weather events.

As summarized above, this study presents a review of industry 4.0 technologies in aviation sector. The technologies discussed in this study are, of course, the tip of the iceberg for the aviation industry in the digital transformation processes. There are also a range of technologies not covered in the study that are already in use in the aviation industry, or are just beginning to be used. These play a decisive role in the future of the sector and are also effective in terms of market conditions. Future studies could explore these areas.

REFERENCES

Adamopoulou, E., & Daskalakis, E. (2023). Applications and technologies of big data in the aerospace domain. *Electronics*, 12(10), 2225.

AIAA. (2023). Aerospace R&D. 12.08.2023. https://www.aiaa.org/domains/aerospaceRandD

Airbus. (2019). IoT: Aerospace's great new connector: The Internet of Things is making its way into the aircraft, its cabin—And beyond. 10.08.2023. https://www.airbus.com/en/newsroom/stories/2019-07-iot-aerospaces-great-new-connector

Airbus. (2023). Autonomous & connected towards safer, more efficient & interoperable flight. 20.10.2023. https://www.airbus.com/en/innovation/autonomous-connected

Airport World. (2022). Robots at the ready at Belfast International Airport. 20.10.2023. https://airport-world.com/robots-at-the-ready-at-belfast-international-airport/

Akkaya, M., & Kaya, H. (2019). Innovative and smart technologies in logistics. In *17th International Logistics and Supply Chain Congress* (pp. 97–105).

Alani, M. M., & Alloghani, M. (2019). Security challenges in the Industry 4.0 era. In *Industry 4.0 and Engineering for a Sustainable Future* (pp. 117–136) Springer, Cham.

Alexandru, A., & Coardoş, D. (2017). Big Data-Concepte, Arhitecturi Şi Tehnologii. *Romanian Journal of Information Technology & Automatic Control/Revista Română de Informatică şi Automatică*, 27(1), 15–24.

Anandan, T. M. (2016). Aerospace manufacturing on board with robots. 13.08.2023. https://www.automate.org/industry-insights/aerospace-manufacturing-on-board-with-robots

Aubert-Tarby, C., Escobar, O. R., & Rayna, T. (2018). The impact of technological change on employment: The case of press digitisation. *Technological Forecasting and Social Change*, 128, 36–45.

AviationHunt. (2023). Explore career options in aviation industry. 12.08.2023. https://www.aviationhunt.com/aviation-careers/

Ayvaz, S., & Alpay, K. (2021). Predictive maintenance system for production lines in manufacturing: A machine learning approach using IoT data in real-time. *Expert Systems with Applications*, 173, 114598.

Badea, V. E., Zamfiroiu, A., & Boncea, R. (2018). Big data in the aerospace industry. *Informatica Economica*, 22(1), 17–24.

Bag, S., Dhamija, P., Singh, R. K., Rahman, M. S., & Sreedharan, V. R. (2023). Big data analytics and artificial intelligence technologies based collaborative platform empowering absorptive capacity in health care supply chain: An empirical study. *Journal of Business Research*, 154, 113315.

Banks, M. (2021). Emerging applications for robotics in the aerospace industry: 5 innovations. https://www.roboticstomorrow.com/story/2021/11/emerging-applications-for-robotics-in-the-aerospace-industry-5-innovations/17755/

Beier, G., Matthess, M., Shuttleworth, L., Guan, T., Grudzien, D. I. D. O. P., Xue, B., ... Chen, L. (2022). Implications of Industry 4.0 on industrial employment: A comparative survey from Brazilian, Chinese, and German practitioners. *Technology in Society*, 70, 102028.

Berman, B. (2012). 3-D printing: The new industrial revolution. *Business Horizons*, 55(2), 155–162.

BigData4ATM. (2016). Big data analytics for socioeconomic and behavioural research in ATM. 20.10.2023. http://www.nommon-files.es/working_papers/BigData4ATM_WhitePaper_May_2016.pdf

Bogue, R. (2018). The growing use of robots by the aerospace industry. *Industrial Robot: An International Journal, 45*(6), 705–709.

Borowski, P. F. (2021). Digitization, digital twins, blockchain, and Industry 4.0 as elements of management process in enterprises in the energy sector. *Energies, 14*(7), 1885.

Boserup, E. (1981). *Population and Technology* (Vol. 255). Oxford: Blackwell.

Boston Consulting Group (BCG). (2019). Embracing Industry 4.0 and rediscovering growth. https://www.bcg.com/capabilities/operations/embracing-Industry-4.0-rediscovering-growth.aspx

Brownlow, L. (2022). How the Internet of Things (IoT) is improving aviation. 12.08.2023. https://worldaviationfestival.com/blog/travel-technology/how-the-internet-of-things-iot-is-improving-aviation/

Buntak, K., Kovačić, M., & Mutavdžija, M. (2019). Internet of things and smart warehouses as the future of logistics. *Tehnički glasnik, 13*(3), 248–253.

Butt, J. (2020). A strategic roadmap for the manufacturing industry to implement Industry 4.0. *Designs, 4*(2), 11.

CEPAL. (2021). Digital technologies for a new future. 20.08.2023. https://www.cepal.org/sites/default/files/publication/files/46817/S2000960_en.pdf

Charoenporn, P. (2018). Smart logistic system by IOT technology. In *Proceedings of the 6th International Conference on Information and Education Technology* (pp. 149–153).

China, C. R. (2023). Smart manufacturing technology is transforming mass production. 20.08.2023. https://www.ibm.com/blog/smart-manufacturing/

Cyert, R. M., & Mowery, D. C. (1987). *Technology and Employment.* Washington, DC: National Academy of Sciences.

Das, M., Sahu, S., & Parhi, D. R. (2021). Composite materials and their damage detection using AI techniques for aerospace application: A brief review. *Materials Today: Proceedings, 44*, 955–960.

Delvadiya, D. (2023). Why is cloud technology so popular in the aerospace industry?. 19.08.2023. https://www.tutorialspoint.com/why-is-cloud-technology-so-popular-in-the-aerospace-industry#:~:text=Data%20Management%20and%20Processing,and%20increased%20productivity%20and%20safety

Dey, S. (2023). How are aerospace robotics redefining innovation in aerospace engineering?. https://swisscognitive.ch/2023/05/15/how-are-aerospace-robotics-redefining-innovation-in-aerospace-engineering/

Ding, B., Ferras Hernandez, X., & Agell Jane, N. (2023). Combining lean and agile manufacturing competitive advantages through Industry 4.0 technologies: An integrative approach. *Production Planning & Control, 34*(5), 442–458.

Dou, X. (2020). Big data and smart aviation information management system. *Cogent Business & Management, 7*(1), 1766736.

Duda, T., & Raghavan, L. V. (2016). 3D metal printing technology. *IFAC-PapersOnLine, 49*(29), 103–110.

DXC. (2023). The future of AI in the aerospace industry. 12.08.2023. https://dxc.com/sg/en/insights/perspectives/paper/the-future-of-ai-in-the-aerospace-industry

EASA, (2023). ARTIFICIAL INTELLIGENCE ROADMAP 2.0 Human-centric approach to AI in aviation. 17.10.2023. https://www.easa.europa.eu/en/downloads/137919/en

Enyoghasi, C., & Badurdeen, F. (2021). Industry 4.0 for sustainable manufacturing: Opportunities at the product, process, and system levels. *Resources, Conservation and Recycling, 166*, 105362.

European Commission. (2017). Industry 4.0 in aeronautics: IoT applications. 25.08.2023. https://ati.ec.europa.eu/sites/default/files/2020-06/Industry%204.0%20in%20Aeronautics%20-%20IoT%20applications%20%28v1%29.pdf

FAA. (2022). Research and development. 10.08.2023. https://www.faa.gov/about/office_org/ headquarters_offices/apl/aee/research

Farooq, M. S., Riaz, S., Abid, A., Abid, K., & Naeem, M. A. (2019). A survey on the role of IoT in agriculture for the implementation of smart farming. *IEEE Access*, 7, 156237–156271.

Fernandes, A., & Ventura, R. (2020). Human-robot collaboration in microgravity: The object handover problem. arXiv preprint arXiv:2005.05735.

Fetters-Walp, E. (2011). Laser technology has far-reaching applications at Boeing, from jetliner production to making windtunnel models. *Boeing Frontiers*, 25, 21–29.

Frank, A. G., Dalenogare, L. S., & Ayala, N. F. (2019). Industry 4.0 technologies: Implementation patterns in manufacturing companies. *International Journal of Production Economics*, *210*, 15–26.

Fraser, J. (2019). The robot airport workers of Seoul. 20.10.2023. https://jordanfraser.medium.com/the-robot-airport-workers-of-seoul-a08349f4f44e

General Electrics. (2023). A variety of additive manufacturing processes. 12.08.2023. https://www.ge.com/additive/additive-manufacturing

Ghobakhloo, M. (2020). Industry 4.0, digitization, and opportunities for sustainability. *Journal of Cleaner Production*, 252, 119869.

Gibson, I., Rosen, D., & Stucker, B. (2014). *Additive Manufacturing Technologies, 3D Printing, Rapid Prototyping, and Direct Digital Manufacturing*. New York: Springer.

Graham, A. (2023). AI in aviation maintenance: How it's changing the industry. 12.08.2023. https://www.linkedin.com/pulse/ai-aviation-maintenance-how-its-changing-industry-andy-graham/

Guerrero-Ibáñez, J., Zeadally, S., & Contreras-Castillo, J. (2018). Sensor technologies for intelligent transportation systems. *Sensors*, *18*(4), 1212.

Gupta, R., & Rathore, R. (2012). *Navigating the Clouds Aviation Industry*. HCL Technologies, February.

Haleem, A., Javaid, M., Qadri, M. A., & Suman, R. (2022). Understanding the role of digital technologies in education: A review. *Sustainable Operations and Computers*, *3*, 275–285.

Heiets, I., & Yibing, X. I. E. (2021). The impact of the COVID-19 pandemic on the aviation industry. *Journal of Aviation*, *5*(2), 111–126.

Hofmann, E., & Rüsch, M. (2017). Industry 4.0 and the current status as well as future prospects on logistics. *Computers in Industry*, *89*, 23–34.

Hussain, A., Hanley, T., Rutgers, V., & Sniderman, B. (2019). Aerospace & Defense 4.0 capturing the value of Industry 4.0 technologies. 15.08.2023. https://www2.deloitte.com/us/en/insights/focus/industry-4-0/aerospace-defense-companies-digital-transformation.html

Ibarra, D., Ganzarain, J., & Igartua, J. I. (2018). Business model innovation through Industry 4.0: A review. *Procedia Manufacturing*, 22, 4–10.

IOTSWC. (2023). How the IoT is improving the aviation industry. 15.08.2023. https://www.iotsworldcongress.com/how-the-iot-is-improving-the-aviation-industry/

Iqbal, S. M., Mahgoub, I., Du, E., Leavitt, M. A., & Asghar, W. (2021). Advances in healthcare wearable devices. *NPJ Flexible Electronics*, *5*(1), 9.

ITU. (2014). The aviation cloud. 12.08.2023. https://www.icao.int/Metings/GTM/Documents/ITU.pdf

Javaid, M., Haleem, A., Vaishya, R., Bahl, S., Suman, R., & Vaish, A. (2020). Industry 4.0 technologies and their applications in fighting COVID-19 pandemic. *Diabetes & Metabolic Syndrome: Clinical Research & Reviews*, *14*(4), 419–422.

Javaid, M., Haleem, A., Singh, R. P., Suman, R., & Gonzalez, E. S. (2022). Understanding the adoption of Industry 4.0 technologies in improving environmental sustainability. *Sustainable Operations and Computers*, *3*, 203–217.

Joshi, S. C., & Sheikh, A. A. (2015). 3D printing in aerospace and its long-term sustainability. *Virtual and Physical Prototyping*, *10*(4), 175–185.

Joshua, E. S. N., Bhattacharyya, D., & Rao, N. T. (2022). The use of digital technologies in the response to SARS-2 CoV2-19 in the public health sector. In P. Ordonez de Pablos, K. T. Chui, M. D. Lytras (Eds.), *Digital Innovation for Healthcare in COVID-19 Pandemic* (pp. 391–418). Academic Press.

Jouav. (2023). The ultimate guide to autonomous drones: Benefits, applications, and top models. 16.08.2023. https://www.jouav.com/blog/autonomous-drones.html

Kanawaday, A., & Sane, A. (2017). Machine learning for predictive maintenance of industrial machines using IoT sensor data. In *2017 8th IEEE International Conference on Software Engineering and Service Science (ICSESS)* (pp. 87–90). IEEE.

Karaoğlu, U., Mbah, O., & Zeeshan, Q. (2023). Applications of machine learning in aircraft maintenance. *Journal of Engineering Management and Systems Engineering*, 2(1), 76–95.

Kell, J. (2023). How the airline industry is using A.I. to improve the entire experience of flying. https://fortune.com/2023/01/31/tech-forward-everyday-ai-airline-industry-fuel-consumption-food-waste/

Koilo, V. (2019). Evidence of the Environmental Kuznets Curve: Unleashing the opportunity of Industry 4.0 in emerging economies. *Journal of Risk and Financial Management*, 12(3), 122.

Kumar, J. A., & Srinivas, G. (2019). Recent trends in Robots Smart Material and its application in aeronautical and aerospace industries. In *Journal of Physics: Conference Series* (Vol. 1172, No. 1, p. 012035). IOP Publishing.

Lee, I., & Lee, K. (2015). The Internet of Things (IoT): Applications, investments, and challenges for enterprises. *Business Horizons*, 58(4), 431–440.

Li, G., Hou, Y., & Wu, A. (2017). Fourth industrial revolution: Technological drivers, impacts and coping methods. *Chinese Geographical Science*, 27(4), 626–637.

Liu, Y., Ma, X., Shu, L., Hancke, G. P., & Abu-Mahfouz, A. M. (2020). From Industry 4.0 to Agriculture 4.0: Current status, enabling technologies, and research challenges. *IEEE Transactions on Industrial Informatics*, 17(6), 4322–4334.

Lohse, K. (2023). The growing role of cloud computing in aerospace. 12.08.2023. https://www.technossus.com/the-growing-role-of-cloud-computing-in-aerospace/

Mariani, J., Zmud, J., Krimmel, E., Sen, R., & Miller, M. (2019). Flying smarter: The smart airport and the Internet of Things. 12.08.2023. https://www2.deloitte.com/za/en/insights/industry/public-sector/iot-in-smart-airports.html

Market Reports World. (2023). Artificial intelligence and robotics in aerospace and defense market drivers, key regions, and applications 2023–2030. 20.10.2023. https://www.linkedin.com/pulse/artificial-intelligence-robotics-aerospace-defense/

McKinney, B. (2023). Using automation and robotics in advanced aircraft production. 17.10.2023. https://www.northropgrumman.com/what-we-do/digital-transformation/using-automation-and-robotics-in-advanced-aircraft-production

Mckinsey, (2022). What are Industry 4.0, the Fourth Industrial Revolution, and 4IR? 24.08.2023. https://www.mckinsey.com/featured-insights/mckinsey-explainers/what-are-industry-4-0-the-fourth-industrial-revolution-and-4ir

Mohamed, M. (2018). Challenges and benefits of Industry 4.0: An overview. *International Journal of Supply and Operations Management*, 5(3), 256–265.

Mujica, L. E., Vehí, J., Staszewski, W., & Worden, K. (2008). Impact damage detection in aircraft composites using knowledge-based reasoning. *Structural Health Monitoring*, 7(3), 215–230.

Müller, J. M., Kiel, D., & Voigt, K. I. (2018). What drives the implementation of Industry 4.0? The role of opportunities and challenges in the context of sustainability. *Sustainability*, 10(1), 247.

Murugan, S. (2023). The impact of technology advancements on the future of the aviation industry. 27.02.2023. https://www.linkedin.com/pulse/impact-technology-advancements-future-aviation-industry-murugan/

NCI. (2022). The main sectors of aviation. 28.08.2023. https://nci.edu/2022/11/21/the-main-sectors-of-aviation/

Niggli, M., & Rutzer, C. (2023). Digital technologies, technological improvement rates, and innovations "Made in Switzerland". *Swiss Journal of Economics and Statistics*, *159*(1), 1–31.

OECD. (2000). Science, technology and innovation in the new economy. Policy Brief, OECD, September.

Oerlikon. (2023). Revolutionizing aerospace with additive manufacturing solutions: Metal 3D printing is reshaping aerospace engineering. 28.08.2023. https://www.oerlikon.com/am/en/markets/aerospace/

Osborne, B., Bryars, R., Spaulding, S., & Pressnell, A. (2022). How are virtual & augmented reality used in aviation training?. 21.08.2023. https://www.higherechelon.com/how-are-virtual-augmented-reality-used-in-aviation-training-2/

Owen-Hill, A. (2023). Robotics in aerospace: What's possible for aerospace companies. 12.08.2023. https://robodk.com/blog/robotics-in-aerospace/

Panwar, N. L., Kaushik, S. C., & Kothari, S. (2011). Role of renewable energy sources in environmental protection: A review. *Renewable and Sustainable Energy Reviews*, *15*(3), 1513–1524.

Paul, M., Maglaras, L., Ferrag, M. A., & AlMomani, I. (2023). Digitization of healthcare sector: A study on privacy and security concerns. *ICT Express*. 9(4), 571–588.

Pietrewicz, L. (2019). Technology, business models and competitive advantage in the age of Industry 4.0. *Problemy Zarządzania*, *17*(2 (82)), 32–52.

Prinsloo, J., Vosloo, J. C., & Mathews, E. H. (2019). Towards Industry 4.0: A roadmap for the South African heavy industry sector. *South African Journal of Industrial Engineering*, *30*(3), 174–186.

Purcell, B. M. (2013). Big data using cloud computing. *Journal of Technology Research*, *5*(1), 1–8.

Ramalingam, T., Christophe, B., & Samuel, F. W. (2017). Assessing the potential of IoT in aerospace. In *Digital Nations–Smart Cities, Innovation, and Sustainability: 16th IFIP WG 6.11 Conference on e-Business, e-Services, and e-Society, I3E 2017, Delhi, India, November 21–23, 2017, Proceedings 16* (pp. 107–121). Springer International Publishing.

Rao, T. R., Mitra, P., Bhatt, R., & Goswami, A. (2018). The big data system, components, tools, and technologies: A survey. *Knowledge and Information Systems*, *60*, 1165–1245.

Ratnasingam, J., Ab Latib, H., Yi, L., Liat, L., & Khoo, A. (2019). Extent of automation and the readiness for Industry 4.0 among Malaysian furniture manufacturers. *BioResources*, *14*(3), 7095–7110.

Renner, T. (2019). Robots bring aeroplane production up to speed. 20.10.2023. https://www.techbriefs.com/component/content/article/tb/supplements/md/features/applications/33714

Rodrigues, D., Carvalho, P., Lima, S. R., Lima, E., & Lopes, N. V. (2022). An IoT platform for production monitoring in the aerospace manufacturing industry. *Journal of Cleaner Production*, *368*, 133264.

Rogers, S. (2019). The role of technology in the evolution of communication. 12.08.2023. https://www.forbes.com/sites/solrogers/2019/10/15/the-role-of-technology-in-the-evolution-of-communication/?sh=77f34c3c493b

Saenz, J. (2015). Robots: Humans' dependable helpers. 20.10.2023. https://www.iff.fraunhofer.de/en/business-units/robotic-systems/valeri.html

Safi, M., & Chung, J. (2023). Augmented reality uses and applications in aerospace and aviation. In A. Y. C. Nee, S. K. Ong (Eds), *Springer Handbook of Augmented Reality* (pp. 473–494). Cham: Springer International Publishing.

Sajid, H. (2023). AI in robotics: 6 groundbreaking applications. 20.10.2023. https://www.v7labs.com/blog/ai-in-robotics#h6

Santos, B. P., Charrua-Santos, F., & Lima, T. M. (2018). Industry 4.0: An overview. In *Proceedings of the World Congress on Engineering* (Vol. 2, pp. 4–6). London, UK: IAEN.

SESAR. (2015). Data driven aircraft trajectory prediction research. 20.10.2023. https://www.sesarju.eu/sites/default/files/documents/awards2021/DART%20CONTRIBUTIONS%20TO%20THE%20DIGITAL%20EUROPEAN%20SKY.pdf

SESAR. (2016). MALORCA. 20.10.2023. https://www.sesarju.eu/projects/malorca

SESAR. (2018). INTUIT – Interactive toolset for understanding trade-offs in ATM performance. 20.10.2023. http://www.nommon-files.es/h2020_project_deliverables/INTUIT_Project_Summary.pdf

SESAR. (2023). Time based separation. 20.10.2023. https://www.sesarju.eu/sesar-solutions/time-based-separation#:~:text=SESAR's%20time%2Dbased%20separation%20(TBS,to%20maintain%20runway%20approach%20capacity)

Sharma, M., Gupta, A., Gupta, S. K., Alsamhi, S. H., & Shvetsov, A. V. (2021). Survey on unmanned aerial vehicle for mars exploration: Deployment use case. *Drones*, *6*, 4.

Sheng, R. (2022). *3D Printing: A Revolutionary Process for Industry Applications*. Cambridge-shire, England: Woodhead Publishing.

Simitzis, P., Tzanidakis, C., Tzamaloukas, O., & Sossidou, E. (2021). Contribution of precision livestock farming systems to the improvement of welfare status and productivity of dairy animals. *Dairy*, *3*(1), 12–28.

Singh, A., Meshram, S., Gujar, T., & Wankhede, P. R. (2016). Baggage tracing and handling system using RFID and IoT for airports. In *2016 International Conference on Computing, Analytics and Security Trends (CAST)* (pp. 466–470). IEEE.

Spatial Team. (2021). Additive manufacturing in aerospace. 12.08.2023. https://blog.spatial.com/additive-manufacturing-aerospace/

Spencer, O. (2022). Technological advancements shaping the future of aviation. 17.08.2023. https://www.linkedin.com/pulse/technological-advancements-shaping-future-aviation-spencer-i-/

Spiegel, R. (2022). Boeing and Northrop Grumman commit to additive manufacturing initiative. 17.08.2023. https://www.designnews.com/3dp/boeing-and-northrop-grumman-commit-additive-manufacturing-initiative

Stewart, M. (2021). How IoT is all set to revolutionize the aviation sector to make it even better. mytechmag.com/how-iot-is-all-set-to-revolutionize-the-aviation-sector-to-make-it-even-better/

Tan, M. & Yang, C. (2023). New air traffic management system to share flight path data aims to save time, fuel in flying. 20.10.2023. https://www.channelnewsasia.com/singapore/new-air-traffic-management-system-share-flight-path-data-aims-save-time-fuel-flying-3564151

Turbide, D. (2017). How are IoT sensors helping the aviation industry? 18.08.2023. https://www.techtarget.com/searcherp/answer/How-are-IoT-sensors-helping-the-aviation-industry

Unhelkar, B., Joshi, S., Sharma, M., Prakash, S., Mani, A. K., & Prasad, M. (2022). Enhancing supply chain performance using RFID technology and decision support systems in the industry 4.0 – A systematic literature review. *International Journal of Information Management Data Insights*, *2*(2), 100084.

Vagdevi, P., & Guruprasad, H. (2015). A study on cloud computing in aviation and aerospace. *International Journal of Computer Science & Engineering Technology*, *6*(3), 94–98.

Venkateswaran, D. N. (2020). Industry 4.0 solutions – A pathway to use smart technologies/build smart factories. *International Journal of Management (IJM)*, *11*(2), 132–140.

Vishnuraj, S. (2022). Future is sky high for additive manufacturing. 20.08.2023. https://www.magzter.com/tr/stories/Business/Manufacturing-Today/FUTURE-IS-SKY-HIGH-FOR-ADDITIVE-MANUFACTURING

Watt, T. (2023). The role of technology in the future and its impact on society. 12.08.2023. https://timesofindia.indiatimes.com/readersblog/amitosh/the-role-of-technology-in-the-future-and-its-impact-on-society-52565/

Wittmer, A., & Vespermann, J. (2011). The environment of aviation. In A. Wittmer, T. Bieger, R. Müller (Eds), *Aviation Systems: Management of the Integrated Aviation Value Chain* (pp. 39–57), Springer.

Xia, F., Yang, L. T., Wang, L., & Vinel, A. (2012). Internet of Things. *International Journal of Communication Systems, 25*(9), 1101–1102.

Yang, C., & Huang, C. (2023). Natural Language Processing (NLP) in aviation safety: Systematic review of research and outlook into the future. *Aerospace, 10*(7), 600.

Youd, F. (2021). Robot workforce: Airports are developing robot workers to tirelessly serve passengers. 20.10.2023. https://airport.nridigital.com/air_may21/robot_airport_workers

Zambon, I., Cecchini, M., Egidi, G., Saporito, M. G., & Colantoni, A. (2019). Revolution 4.0: Industry vs. agriculture in a future development for SMEs. *Processes, 7*(1), 36.

Zhang, K., & Aslan, A. B. (2021). AI technologies for education: Recent research & future directions. *Computers and Education: Artificial Intelligence, 2*, 100025.

10 Sustainability in Airline Industry
Assessment of Airports

Nurcan Deniz
Eskisehir Osmangazi University, Eskisehir, Türkiye

10.1 INTRODUCTION

Nowadays, "sustainability" is one of the megatrends that shapes all of the world. The generally accepted definition of this term can be found in the report of "Our Common Future" (known as Brundtland Report) published by the World Commission on Environment and Development in 1987. The third part of this report is related with "Sustainable Development" and sustainability is defined as *"to ensure that it meets the needs of the present without compromising the ability of future generations to meet their own needs"* in the 27th item. "Economic", "environmental" and "social" are the three elements of sustainability in the light of triple bottom line proposed by Elkington (1994).

It can be useful to explain the concept of "green" initially, which brings together the "economic" and "environmental" elements. On the one hand, the protection of capital and the prevention of its deterioration are aimed in the economic side (Goodland & Bank, 2002); on the other hand, the principles of resilience, flexibility and interconnectedness of ecosystems need to be cared in the environment side (Morelli, 2011) in the "green". "Green" expands to "sustainability" by adding the "social" element which includes issues about personal and societal.

In this context, 17 Sustainable Development Goals (SDGs) were defined covering all these three elements by United Nations according to the "2030 Agenda for Sustainable Development". The aforementioned goals start with "No Poverty" and ends with "Partnerships for the Goals".

Targets and indicators were also defined related to each SDG. Targets 3.6, 9.1 and 11.2 are the directly related targets with sustainable transport. Furthermore, there are more indirectly related SDGs especially SDG 13 (climate action) and SDG 7 (energy). "Sustainable transport" is identified as a "cross-cutting accelerator" to achieve not only SDGs but also Paris Climate Change Agreement. Enhanced safety, universal access, reduced environmental and climate impact, greater efficiency and improved resilience are the objectives of sustainable transport (UN, 2021). Mobility of persons, goods, and services can be achieved with road, rail, maritime and air transport (Sharif, 2018). Air transport is vital in terms of multimodal transport systems. In addition, economic and social contributions to business, tourism, trade, urban forms and settlement patterns cannot be ignored (Liu, 2018).

DOI: 10.1201/9781003389187-10

Air transportation is one of the causes of climate change, noise and air pollution. In this context, sustainable aviation becomes a prominent issue. Sustainability will gain importance considering the predictions about high number of passengers in the future. It is especially important to handle sustainable aviation in a holistic way. Not only the environmental, but also the social dimension, need to be considered, in addition to the economic. Sustainability is also a priority according to CORSIA (Carbon Offset and Reduction Scheme for International Aviation) established by the industry's global standard-setting body called International Civil Aviation Organization (ICAO) in 2016. There is a connection between Strategic Objectives of ICAO (Safety, Air Navigation Capacity and Efficiency, Security and Facilitation, Economic Development of Air Transport and Environmental Protection) and 15 (except 6 and 14) of the SDGs (Liu, 2018; www.icao.int/about-icao). No Country Left Behind (NCLB), Resource Mobilization Strategy, Strategic Planning, Coordination and Partnerships (SPCP), Aviation Partnerships for Sustainable Development (APSD), and Sustainable Mobility for All (SuM4All) are some of the initiatives of ICAO (sdgs.un.org).

As a first global market-based measure, CORSIA is planned to be implemented in three phases. After the initial pilot phase (2021–2023), the first phase will cover the years between 2024 and 2026. The period for the third phase is between 2027 and 2035. According to the data, 115 States voluntarily intended to participate in CORSIA (1 January 2023) (www.icao.int/environmental-protection). International Air Transport Association (IATA), Sustainable Aviation (SA), The International Coalition for Sustainable Aviation (ICSA), and Air Transport Action Group (ATAG) are the other stakeholders in aviation sector. ICAO also formed a partnership with UN-Habitat (Liu, 2018).

Reducing fuel consumption, using sustainable aviation fuels (SAF), elimination of single use plastics, including local dishes in the menus, and training the passengers, are some of the sustainable implementation examples in the aviation sector.

This chapter handles the sustainable aviation according to both environmental, social, and economic aspects. There will be a section about assessment of airports in the sustainability context. The chapter will conclude with an implementation about assessment of airports with a Multi-Criteria Decision-Making (MCDM) technique and a conclusion.

10.2 ENVIRONMENTAL SUSTAINABILITY OF AVIATION

The goals to reduce environmental impacts of airlines can be categorized in seven themes. To reduce CO_2 emissions and air pollution are two main goals related with the first theme is "emissions". Some of the initiatives listed to reach this goal are common (reduce fuel consumption, introduce new fuel-efficient aircraft, etc.). On the other hand, there are some other initiatives (replace employees' company cars, partnerships with NGOs, etc.). Waste, energy, water, biodiversity, and noise are other themes in this context. Recycle onboard waste, reduce energy use in offices/facilities, wash trucks/equipment with rainwater, reduce discharge from maintenance facilities, involvement in environmental conservation projects, introduce quieter aircraft are some of the initiatives listed under these themes (Cowper-Smith & Grosbois, 2011). The environmental impacts are especially important for local residents in the vicinity of airports and under flight paths (Whitelegg, 2000).

Results of the analyses based on corporate social responsibility (CSR) reports reveals that environmental issues gain more importance than social or economic dimensions (Cowper-Smith & Grosbois, 2011).

10.2.1 Noise

According to the results of a bibliometric-based analysis based on green aviation industry, development trends were classified in three clusters: noise, environmental impacts, and green image. Although noise can be classified in the environmental impact category, it has been evaluated in a separate class due to its importance and the earliest in the timeline (Qiu et al., 2021). Similarly, in the framework of SA "noise" is classified as a separate goal in sustainable aviation goals: "quieter". The other goals are "smarter" (social and economic, natural resources, implementation) and "cleaner" (climate change, local air quality, surface access) (SA Report, 2016).

It is known that the community has opposed airport expansion projects, due to complaints about noise of airports. Noise is an important factor that needs to be considered in aircraft design. ICAO had started to regulate noise in the 1960s and developed aircraft noise certification standards (Qiu et al., 2021). Perceived noise levels and impacts can vary (Budd et al., 2019). Beyond annoyance, high levels of noise can be responsible for hearing impairment, sleep disturbance, stress, mental disorders, task performance, and productivity (Whitelegg, 2000).

According to the SA Noise Road-Map to develop new quieter aircrafts and to modify existing ones to be quitter are first type noise reduction techniques. Besides, there are techniques related to operational initiatives such as continuous descent operations, continuous climb operations, and steeper approaches (displacing runway thresholds, horizontal aircraft noise management) (SA Report, 2015). On the other hand, SA Progress Report (2015) draws attention to the importance of land use planning.

10.2.2 Emissions

Aircraft emissions are another important environmental problem causing climate change. It is known that commercial airlines are responsible for approximately 3% of emissions worldwide (Ragbir et al., 2021) and 13% of the total transport industry (Lee et al., 2018). This contribution is measured by "radiative forcing" and in 2050 it is expected that 5–15% of it will be caused by aviation (Whitelegg, 2000). The sector goal defined by IATA is to reduce CO_2 emissions by 50% by 2050, relative to 2005 levels (Lee et al., 2018).

Although it has the largest share and is one of the most dangerous greenhouse gases (Ragbir et al., 2021), carbon dioxide (CO_2) is not the only emission produced by aircraft engines; also Nitrogen Oxides (NO_x), Sulfur Oxides (SO_x), Volatile Organic Compounds (VOC) and Carbon Monoxide (CO) are also emitted. In addition, water vapour, particulate matter, and other pollutants need to be considered. These emissions cause health problems directly and indirectly. Diseases change according to the exposure levels of these pollutants. For example, low levels CO can impair concentration, whereas high levels of CO can cause nausea. Asthma, coughs, headaches, bronchitis, skin irritation, leukaemia are some of the diseases related with these pollutants (ACI Report, 2021; Whitelegg, 2000).

At this point, it is important to stress that these health-related impacts are related to emissions over 3,000 feet (Qiu et al., 2021). It is also reported that fuel toxicity is ten times higher at an altitude of 10,000 metres than ground level. In addition, according to the changes in ozone concentration, the Northern hemisphere is more problematic (Whitelegg, 2000). Not only aircraft emissions, but also aircraft maintenance activities are the sources of emissions (ACI Report, 2021).

Standards and Recommended Practices (SARPs) are the emission standards developed by ICAO (1980s) (Qiu et al., 2021). "Airport Carbon Accreditation (ACA)" was also launched by Airports Council International (ACI) EUROPE line with ISO 14064 and Greenhouse Gas Protocol in 2009 especially on carbon emissions. Airports are accredited through six levels: Mapping–Reduction–Optimisation–Neutrality–Transformation–Transition in this context. According to categorization of ACA, emissions can be originated from three main sources: Airport controlled sources, purchased electricity, and other sources related to the activities of an airport. Vehicles/ground support equipment belonging to the airport, on-site waste and water management, on-site power generation, firefighting exercises, boilers and furnaces, de-icing substances and refrigerant losses are the emission sources in the first category. Off-site electricity generation for heating, cooling, and lighting is the source of the emissions from purchased electricity. Last but not least, third category of emissions related with flights, aircraft ground movements, auxiliary power unit, 3rd party vehicles/ground support equipment, passenger travel to the airport, staff commute, off-site waste and water management staff business travel, non-road construction vehicles and equipment, de-icing substances, and refrigerant losses (ACI Report, 2021).

SAF—also known as drop-in kerosene alternatives—are introduced as the best short-term and medium-term choice, and hydrogen and battery electric are the long-term choice (Cabrera & Melo de Sousa, 2022) to cope with the emission problems. Although there are different definitions of SAF, ICAO defines it as "Alternative Aviation Fuels (AAF) that meet defined sustainability criteria". Whilst first-generation feedstocks (edible crops) origin biofuels are not considered viable alternatives; non-food crops and other alternative raw materials origin second-generation biofuels are more sustainable. On the other hand, algae origin third-generation biofuels are seen as the best choice (Cabrera & Melo de Sousa, 2022).

10.2.3 WASTE

There is a wide variety of waste in airports. The source of these wastes can be shops, passengers, and restaurants. The wastes in this context include general rubbish, food, paper, packaging, newspapers, and plastic bottles. Only passengers are responsible for 6 million tons of waste annually. At this point, Tjahjono et al. (2023) proposed Circular Airport Retail Waste Management (CAWM) based on 9R framework. As the name suggests, it consists of nine waste principles starting from the most circular to most linear. 9R can be categorized in three main classes: Smarter Product Use and Manufacture (Refuse, Rethink, Reduce, and Reuse), Extend Lifespan of

Product (Repair, Refurbish, Remanufacture, and Repurpose) and Useful Application of Materials (Recycle and Recover) (Tjahjono et al., 2023).

10.2.4 TECHNOLOGICAL DEVELOPMENTS

It is important to design more fuel-efficient aircraft (Budd et al., 2019) and the engine is significantly important. Weight reduction, composite materials usage, smaller wings, reducing aircraft painting, are some of the alternatives to reach fuel efficiency (Qiu et al., 2021). Flying-V developed by students at Delft Technical University not only differs from others with using jet fuel or hydrogen but also the passenger cabin, fuel tanks and cargo deck will be placed inside the wing (turkishairlines.com).

10.2.5 AIR TRAFFIC MANAGEMENT AND OPERATIONAL PROCEDURES

Reduction in fuel consumption is not only be possible via technologic developments in aircraft engine or fuels. Route network optimization is another solution to cope with this problem. In addition, operation costs can be reduced in line with shorter flight distances. Airlines are encouraged to conduct cost–benefit analyses and establish performance appraisal methods to improve operational efficiency (Qiu et al., 2021). To enhance air traffic management techniques (continuous climb departures, precision area navigation etc.) (Budd et al., 2019) and to improve maintenance schedules (Cabrera & Melo de Sousa, 2022) are the other activities in this context.

10.3 SOCIAL SUSTAINABILITY OF AVIATION

Environmental sustainability is highly related with the social and economic sustainability. As stated earlier, environmental problems have negative effects on health, culture, wellbeing, and livelihoods (Qiu et al., 2021).

There are some CSR initiatives by airlines within the social dimension. Employee wellbeing and engagement, community wellbeing, diversity, and social equity are themes related to social dimension. To provide medical check-ups and education programmes about the environment for employees are some activities in the employee wellbeing and engagement context. Initiatives are generally related to women and disabled persons to increase diversity. Airlines prefer support educations for young people and donate to charities to increase community wellbeing (Cowper-Smith & Grosbois, 2011).

CSR performance is important both for consumer satisfaction and employee motivation (Wang et al., 2015). Individuals are both potential passengers and senders and buyers of air cargo (Qiu et al., 2021). Airline passengers as well as flight attendants need to be analyzed in the social side in the aviation sector, taking into account that they are the internal stakeholders. There are also studies conducted to discover perceptions of flight attendants about corporate social responsibility (CSR) practices (Ilkhanizadeh & Karatepe, 2017).

The results of the research on Willingness To Pay (WTP) for greener aviation is also important in social sustainability. Happiness and caring were found as significant mediators to explain WTP (Ragbir et al., 2021). Green image is also an important issue for airline reputation regarding purchasing behaviour of green consumers is affected by green image (Qiu et al., 2021).

ICAO Gender Equality Programme (sdgs.un.org) is also a good example in this context. To improve safety (Qiu et al., 2021), is another crucial point, and due to the digital transformation, it is be possible to design incident reporting systems using blockchain to ensure safety in a transparent and accountable way (Aktas et al., 2022).

Beyond "human", it is highly important to analyze "society" in social sustainability. At this point it is important to stress the role of cooperation in paving the way (Qiu et al., 2021). All stakeholders need to come together to reach SDGs. Government plays a role, but public support is also important.

10.4 ECONOMIC SUSTAINABILITY OF AVIATION

As one of the fastest growing and most highly competitive sectors in the economy, there is a trade-off between social and economic benefits and environmental damages in aviation sector (Whitelegg, 2000). On the other hand, initiatives to reduce environmental damages are expensive. For example, SAF usage causes 10–20% price increase in long-distance economy flight ticket (UN, 2021).

The importance of airports is more meaningful with urbanization trends (Liu, 2018) and increased connectivity requirement (Sharif, 2018). The aviation sector also has relationships with other economic sectors such as tourism etc. (Cowper-Smith & Grosbois, 2011).

Airport expansion related employment can be classified in three groups: Direct, indirect and induced. Direct employment related to aviation services must be located at airports. On the other hand, indirect employment is related to the firms involved in aviation (Whitelegg, 2000).

Economic prosperity is the theme related with economic dimension in CSR reports of airlines. Airlines implement a supplier code of conduct to improve procurement actions and use apprenticeships or internship programmes to create jobs (Cowper-Smith & Grosbois, 2011).

10.5 ASSESSMENT OF AIRPORTS BASED ON SUSTAINABILITY CRITERIA

Airports are vital for the promotion of both socio-economic development and environmental pollution (Ramakrishnan et al., 2022), and their assessment is important to reach the SDGs. In 2018, it was reported that the number of airports was nearly 14,000. In addition, new airports were under construction. Airports are unique and complex infrastructures in which there are multiple space types like logistic centres, hotels, and conference facilities. Airport operations are also responsible for emissions (15%) related to the aviation sector. There is also a service infrastructure layer besides the physical infrastructure layer (Ramakrishnan et al., 2022).

A number of tools exist for rating the sustainability of buildings. Leadership in Energy and Environmental Design (LEED) and The Building Research Establishment's Environmental Assessment Method (BREEAM) are categorized in Green Building Rating Tools (GBRT), which are focused on environmental protection.

Dow Jones and ESG (Environmental, Social, Governance) are tools classified in Sustainability Rating Tools (SRT), which considers all three dimensions. According to the sustainability reports of 26 airports between 2002 and 2020 in GRI database, LEED is the most mentioned (50%) GBRT. Airports also stated the participation in the ACA programme (85%). ISO 14001 (58%) and ISO 50001 (27%) are the certifications reported in the context of environmental and energy management systems respectively (Ramakrishnan et al., 2022).

Frameworks in the literature exist to assess green/sustainable airports. In one of these studies Ramakrishnan et al. (2022) proposed a novel green assessment framework for airports. This framework consists of three main layers: airport infrastructure, sustainability dimension, and life-cycle assessment. Physical (control tower, terminals, carparks, etc.) and service infrastructures are the two dimensions of the first layer (airport infrastructure). Sustainability dimensions are consistent with triple bottom line: Economic performance, social equity, and environmental protection. Finally, design and construction, operation and maintenance and project closure are the elements of the third stage (life-cycle assessment). Eight categories were proposed for Green Airports: Site, environment, material, waste, water, energy, transport, and innovation.

There are also other studies in the literature to assess airports with MCDM techniques. Kılkış and Kılkış (2016) developed a "Sustainability Ranking of Airports Index", and also ranked nine airports with this index. This index consists of 5 dimensions and 25 indicators. "Airport Services and Quality", "Energy Consumption and Generation", "CO_2 Emissions and Mitigation Planning", "Environmental Management and Biodiversity" and "Atmosphere and Low Emission" are the dimensions in this index.

In the other MCDM based assessment, Lu et al. (2018) adapted the Sustainability-Balanced Scorecard (SBSC) for the evaluation of the performance of three international airports in Taiwan. Financial perspective, internal business process, learning and growth, environmental perspective, and social perspective are the dimensions in SBSC. They also used MCDM techniques (DEMATEL, ANP and VIKOR) in their model. The results of the study showed that social perspective has the highest degree of net influence and the most crucial factor is airport image for performance evaluation. Customer service culture, airport image, and aviation education and human resource development are the three criteria under social perspective. Carbon emission reduction and energy conservation, green building practices, preventing and monitoring noise are the criteria under learning and growth which are related to environmental perspective.

In a recent study, Öztürk (2019) evaluated sustainability practices of airports in Türkiye using Analytic Hierarchy Process (AHP). In addition to social, environment and economic dimension, operational dimension was also taken into consideration in this MCDM model in which airport activities and ease of transportation were analyzed. Airport employee satisfaction, passenger satisfaction, and community welfare are the main criteria under the social dimension. Occupational health and safety; places such as gyms, places of worship, kindergartens that employees can benefit from; equal employment opportunities; organizing trainings to increase sustainability awareness; all kinds of equipment and designs that will facilitate the movements of the disabled, elderly, and children, and airport security are some of the sub-criteria included in the social dimension. Within the scope of environmentally friendly construction design, there are criteria for noise, water pollution, and emission reduction; efficient water and energy use; waste and material management, and biodiversity protection. The collection and reuse of wastewater, rain and flood waters, and the cultivation of local plant species are exemplary practices in terms of reducing water demand. Direct and indirect economic contributions of airports were included in the decision-making model. Sustainable behaviour incentives include applying separate pricing for aircraft with low emission and noise levels, and providing discount programmes to employees who come to work by bus, bicycle and shared car (Öztürk, 2019).

10.6 IMPLEMENTATION: MCDM BASED AIRPORT ASSESSMENT

An MCDM based implementation to assess the airports is presented next. MCDM techniques enable handling of conflicting criteria, to make a decision. An MCDM based decision-making starts with defining the criteria and alternatives in the decision matrix. The steps of the implementation are visualized in Figure 10.1.

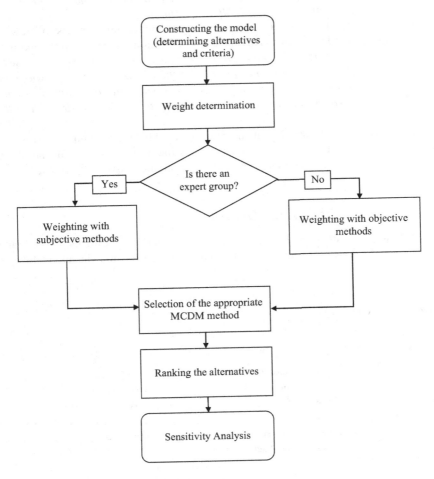

FIGURE 10.1 Workflow of the implementation.

10.6.1 First Step: Constructing the Model

In this step the alternatives and criteria need to be determined. By this way, it is possible to construct the decision matrix which consists of rows (alternatives) and columns (criteria).

The criteria are defined according to the elements of triple bottom line (economic, environmental and social) to assess the airports in a sustainable manner. To refer the aforementioned literature is a good starting point in the previous subtitle. The intention is to select two criteria for each dimension to show the applicability. Using many economic criteria in addition to only one environmental criterion, and one social criterion is inconsistent with sustainability. Table 10.1 denotes the criteria and associated literature. (*Readers can enrich the criteria using literature and real-life examples.*)

There are seven hypothetical airports (alternatives), shown as A, B, C, D, E, F, and G. (*The readers need to check whether an alternative which has the best score in all criteria before the implementation. If there is one apparent alternative, there is no need to compare the alternatives with MCDM techniques.*)

Data (5 Likert) also produced hypothetically with random numbers to show the applicability of the framework. This means all criteria are benefit type (maximization). (*The readers who want to apply this model in a real-world problem, can use the real airports with real data. This means some of the criteria can be cost type (minimization). For example, "CO_2 Emissions" is a cost type criterion which means it is aimed to reduce emissions. To convert all criteria to 5 Likert scale is the other option.*).

The decision matrix starting at the point of an MCDM technique is shown in Table 10.2. The bold and underlined data shows the best alternative for the related criterion. It can be seen that there is not an apparent alternative which is the best for all criteria.

10.6.2 Second Step: Weight Determination

MCDM methods can be used both to determine the weights of dimensions and ranking the airports. Due to the aim of this implementation is to show the applicability of

---

TABLE 10.1
Sustainability Criteria

Sustainability Dimension	Criterion	Literature
Economic	Return on investment (C1)	Lu et al. (2018)
Economic	Non-aviation incomes (C2)	Lu et al. (2018)
Environmental	CO_2 Emissions (C3)	Kılkış and Kılkış (2016)
Environmental	Noise (C4)	Lu et al. (2018) and Öztürk (2019)
Social	Occupational health and safety (C5)	Öztürk (2019)
Social	Facilities for the disabled and the elderly (C6)	Öztürk (2019)

TABLE 10.2
Decision Matrix

	C1	C2	C3	C4	C5	C6
A	1	_5_	3	4	2	_5_
B	3	2	_5_	1	2	3
C	3	4	4	4	4	3
D	1	4	_5_	_5_	_5_	4
E	2	4	2	3	4	3
F	2	1	3	3	2	4
G	_4_	1	2	4	_5_	_5_

MCDM methods in sustainable aviation, only one MCDM technique will be used to rank the airports. Because the steps of the MCDM technique are shown with mathematical expressions primarily in applications, it will be possible to save pages also.

Weights can be calculated by some of MCDM techniques. These techniques mainly classified objective and subjective techniques. Subjective techniques such as AHP depends on expert opinions. Pairwise comparisons can be useful to get the expert decisions. On the other hand, objective techniques such as entropy, use the information from the data. Each weight is determined between $0 < w_j < 1$ and the sum of weights need to be 1.

It was decided to apply equal weights to all criteria. This means each weight of criterion is 0.167 (1/6). The first reason is to demonstrate that MCDM techniques are workable with equal weights if information about weights is not available, and expert knowledge to make pairwise comparison is not readily accessible. In addition, generally economic dimension is more important, but environment and social dimensions are less so in some implementations. By these means, the intention is to focus on all dimensions of sustainability.

10.6.3 THIRD STEP: SELECTION OF THE APPROPRIATE MCDM METHOD

There are many MCDM techniques in the literature (AHP, ANP, TOPSIS, ELECTRE, PROMETHEE, etc.). The intention is to use a relatively new and accessible technique in this chapter.

The Additive Ratio Assessment (ARAS) is an MCDM technique developed by Zavadskas and Turskis in 2010. The details of the technique can be found in the article "A new additive ratio assessment (ARAS) method in multicriteria decision-making". The distinctive feature of ARAS is to add an optimal alternative to decision matrix and to compare the alternatives with the optimal alternative. ARAS consists of five steps as follows (Zavadskas & Turskis, 2010):

Decision-making matrix formation: Decision matrix consists of x_{ij} values (the performance value of the ith alternative in terms of the jth criterion). If the optimal value of j criterion is unknown, the optimal values of the criteria can be determined with equations (10.1) and (10.2) respectively according to the type of criterion (benefit or cost).

$$x_{0j} = \max_i x_{ij}, \text{ if } \max_i x_{ij} \text{ is preferable} \tag{10.1}$$

$$x_{0j} = \min_i x_{ij}, \quad \text{if } \min_i x_{ij} \text{ is preferable} \tag{10.2}$$

Normalization: The second step is normalization of the decision matrix to obtain dimensionless values and it also depends on the type of the criterion. If the criterion is benefit type, $\overline{x_{ij}}$ values in normalized decision matrix ($\overline{X}$) are calculated by equation (10.3).

$$\overline{x_{ij}} = \frac{x_{ij}}{\sum_{i=0}^{m} x_{ij}} \tag{10.3}$$

On the other hand, equations (10.4) and (10.5) need to be used consecutively, if the criterion is cost type.

$$x_{ij}^* = \frac{1}{x_{ij}} \tag{10.4}$$

$$\overline{x_{ij}} = \frac{x_{ij}^*}{\sum_{i=0}^{m} x_{ij}^*} \tag{10.5}$$

Weighting: Weights of each criterion (wj) are used in this step. $\overline{x_{ij}}$ values in normalized decision matrix ($\overline{X}$) are multiplied with the weights (w_j) to obtain a normalized-weighted matrix. Equation (10.6) shows the calculation of $\hat{x}_{ij}$ values in this matrix.

$$\hat{x}_{ij} = \overline{x_{ij}} \times w_j \tag{10.6}$$

Determination of the optimality function values: Equation (10.7) is used to calculate S_i which denotes the optimality function value of ith alternative.

$$S_i = \sum_{j=1}^{n} \hat{x}_{ij} \, i = 0,1,\dots. m \tag{10.7}$$

Utility degree calculation: Lastly, in this step the utility degree (K_i) for each alternative is calculated via equation (10.8). S_0 denoted the optimal function value. K_i shows the relative efficiency and takes value between 0 and 1. The final ranking of the alternatives is produced by setting K_i values in a decreasing order.

$$K_i = \frac{S_i}{S_0} \quad i = 1, 2, \ldots m \tag{10.8}$$

10.6.4 FOURTH STEP: RANKING THE ALTERNATIVES

In this part of the implementation, the aforementioned ARAS steps were followed sequentially. Table 10.3 is the decision matrix with optimal alternative (O). Due to all criteria are benefit type and Likert scale, the data of O alternative are 5. The last row is the total of each column is to be used in the normalization step.

"Normalized decision matrix" and "Weighted normalized decision matrix" are shown in Tables 10.4 and 10.5 respectively.

Finally, Table 10.6 denotes the optimality function values (S_i) and utility degrees (K_i).

TABLE 10.3
Decision Matrix of ARAS

	C1	C2	C3	C4	C5	C6
O	5	5	5	5	5	5
A	1	5	3	4	2	5
B	3	2	5	1	2	3
C	3	4	4	4	4	3
D	1	4	5	5	5	4
E	2	4	2	3	4	3
F	2	1	3	3	2	4
G	4	1	2	4	5	5
Σ	21	26	29	29	29	32

TABLE 10.4
Normalized Decision Matrix

	C1	C2	C3	C4	C5	C6
O	0.24	0.19	0.17	0.17	0.17	0.16
A	0.05	0.19	0.10	0.14	0.07	0.16
B	0.14	0.08	0.17	0.03	0.07	0.09
C	0.14	0.15	0.14	0.14	0.14	0.09
D	0.05	0.15	0.17	0.17	0.17	0.13
E	0.10	0.15	0.07	0.10	0.14	0.09
F	0.10	0.04	0.10	0.10	0.07	0.13
G	0.19	0.04	0.07	0.14	0.17	0.16

TABLE 10.5
Weighted Normalized Decision Matrix

	C1	C2	C3	C4	C5	C6
O	0.04	0.03	0.03	0.03	0.03	0.03
A	0.01	0.03	0.02	0.02	0.01	0.03
B	0.02	0.01	0.03	0.01	0.01	0.02
C	0.02	0.03	0.02	0.02	0.02	0.02
D	0.01	0.03	0.03	0.03	0.03	0.02
E	0.02	0.03	0.01	0.02	0.02	0.02
F	0.02	0.01	0.02	0.02	0.01	0.02
G	0.03	0.01	0.01	0.02	0.03	0.03

TABLE 10.6
**Optimality Function Values
and Utility Degrees**

	S_i	K_i
O	0.18	
A	0.12	0.64
B	0.10	0.53
C	0.13	0.73
D	0.14	0.76
E	0.11	0.59
F	0.09	0.48
G	0.13	0.69

Results show that the best alternative is the D airport with the highest utility degree. The final ranking of airports is D>C>G>A>E>B>F.

10.6.5 FIFTH STEP: SENSITIVITY ANALYSIS

The ranking obtained by MCDM techniques is highly related with the weights of criteria. This means ranking can be affected by changes in the weights. Researchers prefer to change the weights in a range to determine whether the ranking will be affected or not to test the correctness of their decision. It is worth noting that this step is optional.

To show the impact of the weights on the final ranking, it is supposed to give the highest importance to social criteria (0.25). The weights of environmental criteria are assumed to be 0.15, and the weights of economic criteria are 0.10.

Table 10.7 shows the output of ARAS with these new weights.

The final of ranking changed as D>G>C>A>E>B>F. This means the best airport (D) for decision-makers who give greater importance for social dimensions does not

TABLE 10.7

Optimality Function

Values and Utility Degrees

	S_i	K_i
O	0.18	
A	0.12	0.66
B	0.09	0.53
C	0.13	0.73
D	0.15	0.83
E	0.11	0.61
F	0.09	0.53
G	0.14	0.77

change. The only change in the ranking is the mutual exchange between the second and third alternative.

10.7 CONCLUSIONS

Our world has been struggling with environmental and social problems in recent years. Air transportation is one of the causes of climate change, noise, and air pollution. Despite an unexpected intermission provided by COVID-19 (Cabrera & Melo de Sousa, 2022), emission problems soon returned to former levels. In this context, sustainable aviation became an important issue, and will gain importance considering the predictions of increasing passenger levels.

The environmental and social dimensions need to be taken into account, in addition to the economic aspects. It is also especially important to manage sustainable aviation in a multi-disciplinary way. Sustainable aviation pertains to the Department of Tourism and Environment (Cowper-Smith & Grosbois, 2011), the Department of Agricultural Energy (Cremonez et al., 2015) as well as those related to engineering and the built environment (Ramakrishnan et al., 2022). It is also important to bring stakeholders (airlines, government, NGOs, society, etc.) together.

Lean thinking and automation and digitalization have the potential to cope with sustainability related problems (Demir & Paksoy, 2021). Considering the huge impact of human behaviour on waste problems in the airports (Tjahjono et al., 2023), education programmes play a big part in realizing sustainability. Conscious change is important for sustainability, rather than economic forces such as carbon and noise taxes.

Airports were compared in a sustainable way in this chapter to show the applicability of MCDM techniques in sustainable aviation. An additional avenue for researchers is to compare airlines with MCDM techniques. The study of Tanrıverdi et al. (2023) is an example for this kind of study. They used MEREC, CoCoSo, and Borda Voting methods to compare the performance of 56 airlines. The eighth criterion (CO_2 Emissions) is related to the environmental dimension of sustainability. Finally, the fuzzy versions of MCDM techniques are also useful to render more realistic decisions.

REFERENCES

ACI Report (2021) *Airports Council International Airport Carbon Accreditation Annual Report 2019-2021.*

Aktas, E., Demir, S. & Paksoy, T. (2022) The Use of Blockchain in Aviation Safety Reporting Systems: A Framework Proposal, *The International Journal of Aerospace Psychology*, 32:4, 283–306, DOI: 10.1080/24721840.2022.2124161

Brundtland, G. H. (1987) *Report of the World Commission on Environment and Development: Our Common Future* (Report No. A/42/427). United Nations.

Budd, L. C. S., Griggs, S. & Howarth, D. (2019) Sustainable Aviation Futures: Crises, Contested Realities and Prospects for Change, DOI: 10.1108/S2044-9941(2013)0000004013

Cabrera, E. & Melo de Sousa, J. M. (2022) Use of Sustainable Fuels in Aviation—A Review, *Energies*, 15, 2440, DOI: 10.3390/en15072440

Cowper-Smith, A. & Grosbois, D. (2011) The Adoption of Corporate Social Responsibility Practices in the Airline Industry, *Journal of Sustainable Tourism*, 19:1, 59–77, DOI: 10.1080/09669582.2010.498918

Cremonez, P. A. Feroldi, M., Oliveira C. J., Teleken, J. G., Alves, H. J. & Sampaio S. C. (2015) Environmental, Economic and Social Impact of Aviation Biofuel Production in Brazil, *New Biotechnology*, 32:2., 263–271.

Demir, S. & Paksoy, T. (2021) Lean Management Tools in Aviation Industry: New Wine into Old Wineskins, *International Journal of Aeronautics and Astronautics*, 2:3, 77–83.

Elkington, J. (1994) Towards the Sustainable Corporation: Win-Win-Win Business Strategies for Sustainable Development, *California Management Review*, 36:2, 90–100.

Goodland, R. & Bank, W. (2002). Sustainability: Human, Social, Economic and Environmental Social Science, *Encyclopedia of Global Environmental Change*, 6, 220–225.

Ilkhanizadeh, S. & Karatepe O. M. (2017) An Examination of the Consequences of Corporate Social Responsibility in the Airline Industry: Work Engagement, Career Satisfaction, and Voice Behavior, *Journal of Air Transport Management*, 59, 8–17.

Kılkış, Ş. & Kılkış, Ş. (2016) Benchmarking Airports Based on a Sustainability Ranking Index, *Journal of Cleaner Production*, 130, 248–259.

Lee, K. C., Tsai, W. H., Yang, C. H. & Lin, Y. Z. (2018) An MCDM Approach for Selecting Green Aviation Fleet Program Management Strategies under Multi-Resource Limitations, *Journal of Air Transport Management*, 68, 76–85.

Liu, F. (2018) *Promoting Synergy between Cities and Airports For Sustainable Development Report.*

Lu, M. T., Hsu, C. C., Liou, J. H. & Lo, H. W. (2018) A hybrid MCDM and Sustainability-Balanced Scorecard Model to Establish Sustainable Performance Evaluation for International Airports, *Journal of Air Transport Management*, 71, 9–19.

Morelli, J. (2011). Environmental Sustainability: A Definition for Environmental Professionals, *Journal of Environmental Sustainability*, 1:1, 1–9.

Öztürk, N. (2019) *Evaluation of Airports' Sustainability Practices in Turkey Using Analytic Hierarchy Process*, Unpublished MSc Thesis, Istanbul University, Türkiye.

Qiu, R., Hou, S., Chen, X. & Meng, Z. (2021) Green Aviation Industry Sustainable Development towards an Integrated Support System, *Business Strategy and the Environment*, 30, 2441–2452.

Ragbir, N. K., Rice, S., Winter, S. R. & Choy, E. C. (2021) Emotions and Caring Mediate the Relationship between Knowledge of Sustainability and Willingness to Pay for Greener Aviation, *Technology in Society*, 64, 101491.

Ramakrishnan, J., Liu, T., Yu, R., Seshadri, K. & Gou, Z. (2022) Towards Greener Airports: Development of an Assessment Framework by Leveraging Sustainability Reports and Rating Tools, *Environmental Impact Assessment Review*, 93, 106740.

Sharif, M. M. (2018) *Promoting Synergy between Cities and Airports for Sustainable Development Report.*

Tanrıverdi, G., Merkert, R., Karamaşa, Ç. & Asker, V. (2023) Using Multi-Criteria Performance Measurement Models to Evaluate the Financial, Operational and Environmental Sustainability of Airlines, *Journal of Air Transport Management*, 112, 102456.

Tjahjono, M., Ünal, E. & Tran, T. H. (2023) The Circular Economy Transformation of Airports: An Alternative Model for Retail Waste Management, *Sustainability*, 15, 3860.

UN Report (2021) Sustainable Transport, Sustainable Development. *Interagency report for second Global Sustainable Transport Conference.*

Wang, Q., Wu, C. & Sun, Y. (2015) Evaluating Corporate Social Responsibility of Airlines Using Entropy Weight and Grey Relation Analysis, *Journal of Air Transport Management*, 42, 55–62.

Whitelegg, J. (2000) *Aviation: The Social, Economic and Environmental Impact of Flying*, Ashden Trust, London.

Zavadskas, E. K. & Turskis, Z. (2010) A New Additive Ratio Assessment (ARAS) Method in Multicriteria Decision-Making, *Technological and Economic Development*, 16, 159–172.

INTERNET REFERENCES

SA Report (2015) Sustainable Aviation a Decade of Progress 2005-2015, https://www.sustainableaviation.co.uk/ Date of access: 02.11.2022

SA Report (2016) Sustainable Aviation CO_2 Roadmap https://www.sustainableaviation.co.uk/ Date of access: 02.11.2022

https://sdgs.un.org/un-system-sdg-implementation/international-civil-aviation-organization-icao-34579. Date of access: 11.03.2023

https://terminal.turkishairlines.com/en/sustainable-aviation/. Date of access: 29.12.2022

https://www.icao.int/about-icao/aviation-development/Pages/SDG.aspx. Date of access: 04.04.2023

https://www.icao.int/environmental-protection/CORSIA/Pages/state-pairs.aspx. Date of access: 04.04.2023

Index

Pages in *italics* refer to figures and pages in **bold** refer to tables.

Printed in the United States
by Baker & Taylor Publisher Services